*Gases, liquids and solids*

# Gases, liquids and solids

D. TABOR

*SECOND EDITION*

The right of the
University of Cambridge
to print and sell
all manner of books
was granted by
Henry VIII in 1534.
The University has printed
and published continuously
since 1584.

**CAMBRIDGE UNIVERSITY PRESS**

CAMBRIDGE

LONDON · NEW YORK · NEW ROCHELLE

MELBOURNE · SYDNEY

Published by the Press Syndicate of the University of Cambridge
The Pitt Building, Trumpington Street, Cambridge CB2 1RP
32 East 57th Street, New York, NY 10022, USA
10 Stamford Road, Oakleigh, Melbourne 3166, Australia

First edition copyright © David Tabor 1969
Second edition © Cambridge University Press 1979

First published by Penguin Books Ltd 1969
Second edition published by Cambridge University Press 1979
Reprinted 1983, 1985 (with corrections)

Printed in Great Britain at the University Press, Cambridge


*Library of Congress cataloguing in publication data*

Tabor, David.
Gases, liquids, and solids.

Includes index.
1. Gases, Kinetic theory of. 2. Liquids.
3. Solids. I. Title.
OC175.T3 1979   530.4   78-26451

ISBN 0 521 22383 0 hard covers
ISBN 0 521 29466 5 paperback

# Contents

# *Preface to the first edition*

The subdivision of physics into mechanics; heat, light and sound; magnetism and electricity; properties of matter; and kinetic theory, goes back to the early days of classical physics. During the last twenty or thirty years there has been a discernible trend towards a regrouping of subjects, partly to allow for new knowledge and partly to allow for new methods of approach. Few of the areas have received such an impulse as those which, in classical days, were grouped together as kinetic theory and properties of matter. The new approach is to emphasize the atomic structure of matter and to show that, by assuming the existence of attractive and repulsive forces between atoms and molecules and the presence of thermal energy, it is possible to explain nearly all the bulk properties of gases, liquids and solids in terms of relatively simple models.

This book, which attempts to do this, is based on the lecture course given to first-year students at Cambridge. Its treatment is relatively simple and contains very little quantum physics or wave mechanics. It represents an attempt to bridge the gap between sixth-form physics and physical chemistry, and the more advanced courses which follow in later years of specialization. If it has any merit at all this is largely due to the devotion of many generations of Cavendish teachers who have hammered out the type of treatment and approach I have given here. I am particularly indebted to Dr T. E. Faber who first explained the Cavendish syllabus to me, and to Dr D. Shoenberg who has run a parallel course of lectures with me for the last few years and who has, on several occasions, saved me from perpetuating mistaken ideas. I also wish to thank the members of the Cavendish who have made comments on various parts of the manuscript, and Miss Shirley V. King of Birkbeck College for her comments on the first part of Chapter 10. Any mistakes, however, either of fact or of opinion are my own and I accept responsibility for them.

Dr Leo Baeck, in his spirited critique of romantic religion, referred to orthodoxy in the following terms: 'that bent of mind which would cultivate respect for the answer but, in the process, often loses respect for the problem'. A textbook such as this must naturally pay great respect to the answer. The student quite rightly expects to find the correct answer crisply and clearly given. This I have attempted to do. But it sometimes happens in physics, as in other walks of life, that the simple crisp answer is not the whole truth. Where it seems profitable I have discussed this as openly as possible. I hope it does not confuse the student, for this is not my intention. I recognize that, so long as a student's ability is judged by examinations, both student and examiner will persist in cultivating great respect for the answer. On the other hand, the progress of physics, in the long run, may well depend far more on our according greater respect to the problem.

Cambridge 1969

# *Preface to the second edition*

The main change in this edition is the conversion to SI units. I have tried to be as consistent as possible and have been greatly aided in this task by the unflagging and meticulous care of the publishers. I have occasionally retained the °C particularly when referring to historical experiments and to the properties of water or ice near the melting point; 0 °C seems a more sensible point of reference than 273 K.

I have included some fresh material which fills out certain omissions in the first edition. I have also added some new themes which extend the coverage of the book; in particular the nature of van der Waals forces and the basic ideas of colloid stability. In some places, where it seemed relevant, I have added sections at a level rather more advanced than that required by the average readers; they may at some later stage find it useful to refer back to these items. Finally I have, as is inevitable in moving from c.g.s. to SI units, completely rewritten the introductory parts of the chapters dealing with the dielectric and magnetic properties of matter.

It remains for me to thank various colleagues and numerous correspondents who at various times have written to me proposing corrections and improvements. Though I have not been able to implement all their suggestions I am most grateful to them. For if this new edition represents any improvement on the old it is due in no small measure to their involvement. The faults are all my own.

Cambridge 1979

Note added at 1985 reprinting

In this reprint several typographical errors have been corrected and one or two misleading statements rectified. I am grateful to various correspondents who have written to me suggesting such improvements. In particular I wish to thank Dr H. Mykura of Warwick University for a number of detailed and perceptive comments.

Cambridge 1985

# Introduction

The three states of matter are the result of a competition between thermal energy and intermolecular forces. It is this struggle which determines whether a given substance, under given conditions, is a gas, a liquid or a solid. Many textbooks which describe the bulk properties of matter tend to lose sight of this and to treat gases, liquids and solids as though they are quite unrelated materials. Indeed gases become an exercise in kinetic theory, solids an example of the laws of elasticity, and liquids a type of material which shows viscosity and surface tension. In this book we attempt to show that many of the bulk properties of gases, liquids and solids can be explained in terms of intermolecular forces; in this way the three states are linked by a common property. We shall not give a detailed account of the origin of molecular forces but, assuming they exist, we shall show the part they play in determining the properties of matter.

Chapter 1 is a simple account of the nature of atoms and molecules and the forces between them, and includes a rather more detailed explanation of van der Waals forces. The second chapter deals with temperature and the concept of heat. These are the introductory chapters.

The next three chapters describe the properties of gases. This is a well-worn area where the molecular approach has long been used to describe the bulk behaviour. Chapter 3 gives the simple kinetic theory of ideal gases, while Chapter 4 deals with a more sophisticated treatment which includes a simplified approach to the Boltzmann distribution, and a discussion of the equipartition of energy. Chapter 5 deals with imperfect gases where perfect gas behaviour is modified to allow for the finite size of the molecules and for the forces between them.

We turn at once from gases to solids. In gases the molecules are virtually free, in solids they are bound to particular sites and their main thermal exercise consists in vibrating about their equilibrium positions. However, the molecular forces that cause the real gas to deviate from ideal gas behaviour are the same forces that are responsible for the existence of the solid state. In Chapter 6 we show that these forces can explain the heat of sublimation of solids, their thermal expansion, their elastic properties and the existence of the surface energy of solids. Some of these themes are analysed in greater detail in subsequent chapters.

In Chapter 7 the elastic properties of solids are described in terms of inter-molecular forces. This is applied quantitatively to ionic and van der Waals solids and empirically to metals. An account is also given of the elasticity of rubber. Chapter 8 deals with the plastic and brittle properties of solids and shows that these may be understood in molecular or atomic terms. For both types of strength property it is shown that the real strength is generally much less than the theoretical because of the presence of imperfections. Chapter 9 describes the thermal and electrical properties of solids. In this chapter we give a brief account of the specific heat capacities of solids in terms of atomic vibrations (the Einstein and Debye theories); we also show how thermal expansion arises as the amplitude of the vibration increases with increasing temperature. Finally we describe thermal and electrical conductivity; with metals these are both attributed to the transport of energy by the free electrons. The classical theory which treats the electrons as a gas gives the right answers for the wrong reason and it is shown how the theory must be modified to allow for the discrete energy levels of the electrons.

The next two chapters deal with the liquid state, the Cinderella of modern physics. The main difficulty here is that, although in some ways the intermolecular forces dominate (a liquid occupies a specified volume unlike a gas), in other ways thermal motion dominates (a liquid shows mobility, unlike a solid). The molecules are both bound and free so that over small regions they appear to be highly ordered as in a solid whilst over large regions such order does not exist. These features and some recent descriptive ideas (particularly those of Bernal and of Alder) are given in Chapter 10, together with a molecular theory of vapour pressure and surface tension. In Chapter 11 we discuss the flow of liquids and, using the simplest form of Eyring's theory, show that viscosity may be understood in terms of intermolecular forces. Consequently, we are able to show that there is a simple connection between surface tension and viscosity. Finally, before leaving the liquid state, a brief account is given of the factors that determine the stability of colloidal suspensions in a liquid medium.

The last two chapters deal with the dielectric and magnetic properties of matter. It is shown that these properties arise from the polarization of atoms and molecules produced by the applied electrostatic or magnetic field. The dielectric behaviour, described in Chapter 12, receives a very satisfactory explanation in terms of simple atomic and molecular models. In addition, the treatment provides a direct correlation between dielectric properties, optical properties and van der Waals forces. The explanation of magnetic properties is more difficult since quantum effects are of major importance. The classical explanation of diamag-netism which we reproduce in Chapter 13 yields the right answer but, at a more fundamental level, is quite wrong. Paramagnetism is more easily explained and, assuming the existence of strong interactions, a satisfactory and satisfying model for ferromagnetism is found.

Some of the chapters lend themselves to brief summaries and where this has been possible they have been introduced at or near the end of the relevant chapter.

# 1

## Atoms, molecules and the forces between them

### 1.1 Atoms and molecules

#### 1.1.1 *The evidence for atoms and molecules*

Matter is not a continuum of uniform density, but consists of discrete particles or, if one wishes to be more up to date, of localized regions of very high density separated by regions of almost zero density. The particulate nature of matter has been known to us since the time of the Greeks and the idea of atoms (units which could not be further cut or divided) is generally attributed to Democritus (*c.* 460–370 BC). Many of his concepts have a surprisingly modern ring and constitute a tribute to the immense power, as well as the originality, of the Greek approach to logical inference and abstract reasoning. Much of his work was, indeed, the result of thought rather than of direct experiment. One must not, however, read too much modern science into these early ideas. For example in a passage quoted by Theophrastus, *De Sensu*, 61–2, Democritus states, 'Hard is what is dense, and soft what is rare . . . Hard and soft as well as heavy and light are differentiated by the position and arrangement of the voids. Therefore iron is harder and lead heavier.' But it would be wrong from this to attribute to Democritus a knowledge of dislocations and point defects.

The first real attempt to get to grips with the basic 'atoms' of matter had to wait until more quantitative measurements and generalizations had been made. The law of multiple proportions, largely due to the work of J. Dalton (1766–1844) was one of these. It stated that if two elementary substances combined chemically to form more than one compound the weights of one which combine with a fixed weight of the other are in a simple ratio to one another. For example nitrogen and oxygen combine to form nitrous oxide, nitric oxide, nitrogen sesquioxide and nitrogen dioxide; three of these oxides were, in fact, amongst the compounds which Dalton first studied in the course of his investigation. In these compounds 14 unit masses of nitrogen combine respectively with 8, 16, 24 and 32 unit masses of oxygen: i.e. the ratio is $1:2:3:4$. This fits in naturally with the concept of unit masses which can combine in simple multiples with one another.

We consider a simpler case, the reaction of hydrogen and oxygen to form water. Quantitatively we find 1 unit mass of hydrogen combines with 8 unit masses of oxygen to form 9 unit masses of water. According to Dalton's atomic theory (Dalton 1808) we should write

$$1 \text{ atom hydrogen} + 1 \text{ atom oxygen} \rightarrow 1 \text{ atom water}$$

or

$$H + O \rightarrow HO.$$

When these reactions are studied in the gaseous state and the volumes are measured at a standard temperature and pressure, the equation becomes

$$2 \text{ volumes hydrogen} + 1 \text{ volume oxygen} \rightarrow$$
$$2 \text{ volumes water vapour.}$$

The next step in the argument depends on the hypothesis put forward by A. Avogadro in 1811 that at a specified temperature and pressure a given volume of any gas contains the same number of unit masses.

$$2 \text{ unit masses of hydrogen} + 1 \text{ unit mass of oxygen} \rightarrow$$
$$2 \text{ unit masses of water.}$$

This statement makes sense only if the 'unit mass' of oxygen is itself divisible into two equal parts each of which goes into one of the unit masses of water. Arguments along these lines lead finally to chemical equations of the type

$$2H_2 + O_2 \rightarrow 2H_2O,$$

where $H_2$, the smallest quantity of hydrogen which can exist in the free state, is called the molecule whilst H, the smallest quantity of hydrogen to enter into chemical combination, is called the atom. Avogadro's 'unit masses' are therefore molecules.

### 1.1.2 *Avogadro's law*

This is still called a hypothesis but deserves to be promoted to the status of a law. It states that at a specified temperature and pressure a fixed volume of any gas or vapour contains the same number of unit masses, i.e. molecules or atoms if the free molecule is monatomic. If we redefine the atom of hydrogen as 1 unit mass, the molecule of hydrogen consists of 2 unit masses. In the older literature these quantities in grams were called the gram-atom or gram-molecule respectively. In SI units the standard is no longer hydrogen but carbon-12: the gram-atom and gram-molecule are replaced by the mole (and the unit is written 'mol'). It is defined as the 'amount of substance of a system which contains as many elementary particles as there are atoms in 0.012 kg of carbon-12. I can only regard this definition in terms of 12 grams of carbon as a failure of nerve on the part of physical chemists. The logic of SI units demands the use of 12 kg of carbon 12 as the unit and indeed some physicists use the kilomole in this way. We shall not, however, introduce such units into this book. One mole of any gas at a pressure of 760 mm Hg (101 325 Pa) and a temperature of 273 K fills approximately $0.0224 \, \text{m}^3$. The number of molecules in a mole of any substance is called Avogadro's number ($N_A$) and has the value of $6.03 \times 10^{23}$. For example ($N_A$) molecules of hydrogen weigh approximately 2 g; of nitrogen 28 g; of gold (atomic) 197 g and of iodine 254 g: these constitute the mole-mass of these materials to the nearest integer.

### 1.1.3 *First experimental deduction of the size of molecules*

Pioneer work by Fraülein Pockels in 1891 showed that certain organic molecules with polar end groups such as oleic acid which are insoluble in water in bulk will spread over the surface of clean water. Lord Rayleigh experimented further with this and showed in 1899 that if the amount of acid added to the water surface is less than a certain amount there is practically no reduction in surface tension of the water. If, however, the amount added exceeds a certain critical value the surface tension falls rapidly. This is shown schematically in figure 1. Rayleigh assumed that this change occurred when there was a continuous monomolecular layer of fatty acid on the surface. He was thus able to determine the weight of acid per $\text{m}^2$ of surface at which the monolayer was formed. Assuming the monolayer to have the same density as the acid in bulk, the thickness of the monolayer was calculated and found to be about $10^{-9}$ m, i.e. 10 Å. This was the first direct indication that atomic and molecular dimensions are of the order of angstroms.

### 1.1.4 *Determination of Avogadro's number $N_A$.*

We shall mention three very different ways of determining $N_A$. They are all of historic interest but only the last two provide reasonably accurate values.

**1.1.4.1** *Radioactive disintegration.* Radium disintegrates to give off $\alpha$ particles and leave radon gas. Radon disintegrates to give off $\alpha$ particles and leave behind polonium. Further stages in the disintegration are not, in the present context, of great importance. The $\alpha$ particles emitted can be counted by allowing the particles emitted within a small known solid angle to strike a fluorescent screen. Each particle may be counted since it produces an individual scintillation on the screen. From the size of the solid angle chosen the total emission over a solid angle of $4\pi$ can be deduced. Lord Rutherford in 1909 showed that the $\alpha$ particle is a helium atom which has lost its two orbital electrons (see below). These electrons are picked up by the particles so that each $\alpha$ particle becomes a helium atom, and since helium is a monoatomic gas each $\alpha$ particle thus becomes a gas atom. The volume of the helium gas produced in this way can be measured. By extrapolation Rutherford found that, in one year, one kilogram of radium would emit a total of $11.6 \times 10^{20}$ particles which would turn into $43 \times 10^{-6} \, \text{m}^3$ of helium gas at s.t.p.

Consequently $43 \times 10^{-6} \, \text{m}^3$ He contains $11.6 \times 10^{20}$ atoms.
Therefore $22.4 \times 10^{-3} \, \text{m}^3$ He contains $6 \times 10^{23}$ atoms.

This derivation is of interest since it showed for the first time that atoms are individual particles.

Figure 1. Schematic diagram showing the surface tension of water as small quantities of surface-active material are added. The surface tension remains fairly high until a critical amount, corresponding approximately to a complete monolayer of surfactant, has been added; it then falls rapidly to a constant lower value.

amount of surface active material added

1.1.4.2. *Electrolysis.* If a quantity of electricity known as a faraday is used to electrolyse a dilute acid, exactly 1 gram of hydrogen is liberated. Since the faraday is equal to 96 500 coulombs, this means that 96 500 coulombs are involved in converting $N_A$ ions of hydrogen into $N_A$ atoms of hydrogen gas. The charge on the hydrogen ion is due to the loss of a single electron, hence

$$96\,500 \text{ coulomb} = \text{charge of } N_A \text{ electrons} = N_A e.$$

The next step is to determine $e$. This was carried through successfully by R. Millikan in 1909. He produced very fine droplets of oil between the plates of a parallel plate condenser, the plates being set in a horizontal plane. The particles gradually fall because of gravity. If, however, the air surrounding them is subjected to ionizing radiation the particles will pick up charge of say $q$. It then becomes possible to apply an electrostatic field $E$ which is able just to hold the drop in equilibrium in a stationary position. In this state

$$Eq = mg.$$

Therefore as $m$, $g$ and $E$ were known, $q$ was calculated and found always to be an integral number of some fundamental unit of charge. This was assumed to be the charge $e$ of the electron. The current value of $e$ found in this way is very nearly

$$e = 1.6 \times 10^{-19} \text{ coulomb.}$$

Hence

$$N_A = \frac{96\,500}{1.6 \times 10^{-19}} = 6.03 \times 10^{23}.$$

1.1.4.3 *X-rays.* X-rays are electromagnetic waves of short wavelength (about $10^{-10}$ m). M. von Laue was the first to suggest, in 1912, that since the spacing of crystal planes in a simple inorganic crystal is of this order, it ought to be possible to use the crystal as a diffraction grating for X-rays. The wavelengths $\lambda$ of the X-rays may be determined accurately by diffracting them with a conventional ruled grating, at glancing incidence. The X-rays may then be used to find the lattice spacing in the crystal. If, for example, a collimated beam of X-rays strikes a crystal plane at an angle $\theta_1$, it is first refracted through an angle $\theta_2$, but since the refractive index for X-rays is almost unity one may take $\theta_1 = \theta_2 = \theta$. The condition for reinforcement is then that $2d \sin \theta = n\lambda$, where $d$ is the distance between scattering planes and $n$ is an integer. By observing the angles for which strong diffracted beams are obtained, the various scattering planes and their separation may be found. In this way the crystal structure and the lattice spacings may be deduced.

As an example, consider the sodium chloride crystal. This has a cubic structure with sodium and chloride ions at alternate sites (see figure 2). The distance $a$ between each site is found by X-ray analysis to be 0.2818 nm. Consider one of the unit cubes of side $a$. Each of these cubes has 4 sodium and 4 chloride ions, i.e. 4 NaCl molecules. But each corner of the cube is shared by eight contiguous cubes. Consequently each volume $a^3$ contains one-eighth of four molecules of NaCl $=\frac{1}{2}$ molecule of NaCl. If $M$ = molecular weight, $\rho$ = density, the volume occupied by the half-molecule is $\frac{1}{2}(M/N_A)(1/\rho)$. Consequently

$$a^3 = \frac{1}{2}\frac{M}{N_A}\frac{1}{\rho}.$$

Inserting $M = 0.0585$ kg, $\rho = 2170$ kg m$^{-3}$, $a = 0.2818$ nm,

$$N_A = 6.02 \times 10^{23}.$$

### 1.1.5 *Energy states in atoms: the earliest evidence for quantization*

By the turn of the century the work of J. J. Thomson and others had shown that the atom consists of a positively charged massive nucleus and several very light negatively charged particles – the electrons. The problem that perplexed physicists for the next 25 years was the way in which the electrons were linked to the nucleus and the factors that determined the energies of the electrons.

Figure 2. Structure of sodium chloride crystal. The sodium and chloride ions are arranged alternately on a cubic lattice, the separation between ions being 0.2818 nm. The sodium fluoride structure is identical except that $a = 0.231$ nm. The unit cell size is $2a$.

The following argument concerning the energy of electrons within the atom has been given by Sir Neville Mott,† and brings out very simply the need for quantization. Consider the thermal properties of a monatomic gas such as neon. The atom has 10 electrons interacting in some way with the nucleus. Careful calorimetric measurements show that the thermal energy per atom of neon is exactly $\frac{3}{2}kT$ where $k$ is Boltzmann's constant $(1.380 \times 10^{-23} \text{ J K}^{-1})$ and $T$ is the absolute temperature. If we increase the temperature of the gas by 1 degree the energy increase per atom (at constant volume) is exactly $\frac{3}{2}k$. Now the kinetic theory of gases shows that $\frac{3}{2}kT$ is the kinetic energy of a single particle in equilibrium with its surroundings at temperature $T$. This means that the only energy which is changed when we heat a neon atom is its kinetic energy as a single entity. If anything within the atom is moving it is not speeded up by the addition of heat, i.e. no energy has been transferred to the constituent parts of the atom.

This has an important consequence for if the electronic energies were continuous they would have to share in the thermal energy of the atom as a whole. They do not do so. This implies two things. First, the electronic energy states are discontinuous: they are at certain specific levels and no energies between these levels can exist. Secondly the gap between these energy levels must be very large compared with normal thermal energies: for this reason electrons cannot be promoted to higher levels at ordinary temperatures.‡ In fact the gaps between the electron energy levels are of order 5 eV which is very much greater than the thermal energy per atom, about $\frac{1}{40}$ eV at room temperature.

This argument provides, with some hindsight, the simplest and most basic evidence for the existence of quantized energy states within the atom.

### 1.1.6 *Basic structure of the atom*

It is not the purpose of this book to provide a full account of atomic structure. This would involve the wave properties of matter and other concepts used in describing electronic configurations. We shall very briefly fall back on the simple Bohr model: for although this has been superseded by the ideas of orbitals and the uncertainty principle it provides a very convenient model on which to base many order-of-magnitude calculations.

† N. Mott, *Contemporary Physics*, vol. 5 (1964), p. 401.
‡ This has a bearing not only on the specific heat of a gas but also on the Langevin treatment of diamagnetism described on pp. 305–8.

If the nucleus has a charge $Ze$ and a specified electron of mass $m$, charge $-e$, describes a circular orbit of radius $r$ at a speed $v$, we may balance the centrifugal force $mv^2/r$ with the Coulombic attraction between electron and nucleus. Assuming that there is no shielding of charge by other electrons we may write

$$mv^2/r = \frac{Ze^2}{r^2} \frac{1}{4\pi\varepsilon_0}, \tag{1.1}$$

where $\varepsilon_0$ is the permittivity of free space. The kinetic energy of the electron is

$$\tfrac{1}{2}mv^2 = \frac{1}{2} \frac{Ze^2}{r} \frac{1}{4\pi\varepsilon_0}. \tag{1.2}$$

The potential energy is the work done in carrying the electron from infinity to a distance $r$ from the nucleus. This is given by

$$\int_\infty^r \frac{Ze^2}{r^2} \frac{1}{4\pi\varepsilon_0} \, \mathrm{d}r = -\frac{Ze^2}{r} \frac{1}{4\pi\varepsilon_0}. \tag{1.3}$$

Note that attractive forces always give negative potential energies. The total energy $W$ of the electron, which is the sum of the potential and kinetic energies, is the sum of equations (1.2) and (1.3) so that

$$W = -\frac{1}{2} \frac{Ze^2}{r} \frac{1}{4\pi\varepsilon_0}. \tag{1.4}$$

The major defects of this model are twofold. First, an electron rotating about a nucleus should, if treated classically, radiate electromagnetic energy and spiral into the nucleus. Secondly, even if such orbits were stable all possible values of $W$ should be possible corresponding to all possible values of $r$.

When the wave nature of the electron is incorporated into this model both these difficulties disappear. The idea of a crisp orbit is now replaced by a somewhat fuzzy orbital describing the probability of locating the electron. Such an orbital is found to correspond to a stable energy state and no energy is radiated by the electron. Further, a whole series of stable orbitals are permitted as determined by certain (integral) quantum numbers. For example the angular momentum which in the simple Bohr model has a value $mvr$ must be quantized to be $n(h/2\pi)$, where $n$ is an integer and $h$ is Planck's constant ($6.626 \times 10^{-34}$ J s). We may pursue this a little further. We write

$$mvr = n(h/2\pi) \tag{1.5}$$

so that

$$mv^2 = \frac{v}{r} mvr = nh(v/2\pi r). \tag{1.6}$$

But $2\pi r$ is the circumference of the orbit (or the orbital) so that $v/2\pi r$ is the frequency $\nu$ with which the electron orbits the nucleus. Comparing equations (1.4) and (1.2) we see that the total energy $W$ is equal to $-\frac{1}{2}mv^2$. We thus obtain

$$W = -\frac{1}{2}nh\nu = -\frac{1}{2}\frac{nZe^2}{r}\frac{1}{4\pi\varepsilon_0}. \tag{1.7}$$

For the hydrogen atom, where $Z = 1$, the frequency $\nu$ becomes

$$h\nu = \frac{-e^2}{r}\frac{1}{4\pi\varepsilon_0}, \tag{1.8}$$

a result we use later in this chapter in deriving van der Waals forces.

Of course equations (1.7) and (1.8) still contain $r$. For each integral value of $n$ in equation (1.5) there will be an associated radius $r$ and if this is fed into equation (1.4) we shall obtain a whole series of permitted energy levels. The result is

$$W = -\frac{1}{n^2}\frac{Z^2e^4m}{2(4\pi\varepsilon_0)^2}\left(\frac{2\pi}{h}\right)^2. \tag{1.9}$$

Using known values of $e$, $m$ and $h$ we find that for the hydrogen atom ($Z = 1$) the deepest energy level ($n = 1$) has a value of about $24 \times 10^{-19}$ J or 15 eV; the next level ($n = 2$) has a value of 3.75 eV so that there is a gap of over 11 eV. With larger atoms containing more electrons, higher values of $n$ are involved for the lowest energy states but $Z$ is also larger, even if we allow for the effect of screening of the nuclear charge by some of the orbital electrons. In general, the gap between the lowest energy level and the next highest level is of order 5 eV.

Quantum mechanics gives a slightly more complicated picture. The energy levels remain quantized and are still given by equation (1.9), where $n$ is the principal quantum number and can have any value from 1 upwards. The angular momentum, however, is no longer given by $n(h/2\pi)$: it has a value $l^{\frac{1}{2}}(l+1)^{\frac{1}{2}}(h/2\pi)$, where $l$ is an integer which can be either positive or negative but numerically cannot be greater than $n-1$. This has an unexpected consequence. The lowest energy state is $n = 1$: to this corresponds $l = 0$. This means that the lowest energy state (the $S$ orbital) has zero angular momentum – the electron does not move in a circle, it moves in and out radially. There is no simple classical analogue although one might regard the smeared electron cloud as pulsing in and out.

An orbiting electron behaves like a circulating current and its interaction with an applied magnetic field determines the magnetic properties of the atom or molecule. For this reason the quantization of angular momentum is of great impurtance in explaining atomic magnetism. The detailed behaviour is very complicated since the angular momentum must itself follow further quantization rules when exposed to a magnetic field. However, as we shall see in a later chapter, many of the basic magnetic properties of atoms may be understood in terms of the simplest models.

### 1.1.7 *Charge, mass and configuration of atoms: the periodic table*

We may now summarize the main features of atomic structure and establish the basis on which the periodic table of elements may be constructed.

1. An atom consists of a minute nucleus, which is positively charged and contains the main mass of the atom.

2. The nucleus consists of protons and neutrons. These are of equal mass but each proton carried a positive charge whilst the neutron carries no charge. The resultant positive charge (referred to the proton as unit charge) is called the atomic number $Z$, i.e. $Z$ is equal to the number of protons in the nucleus.

3. As a crude approximation, valid especially for smaller atoms of atomic number less than 20, there are about equal numbers of protons and neutrons so that the mass of the nucleus, referred to the proton as unit mass, is about twice the atomic number. This is known as the mass number and in practice is the integer nearest to the mass of the atom, referred to the carbon-12 atom as 12.000. In the older literature the mass number is sometimes loosely referred to as the atomic weight.

4. The nucleus is surrounded by electrons. These have a mass only $\frac{1}{1840}$ of the mass of the H atom (i.e. of a proton or neutron) and they have a unit negative charge.

5. Since the atom is electrically neutral the number of orbital electrons is equal to the atomic number.

6. The chemical properties of the atom are determined by the configuration of the electrons, especially of the outermost ones.

7. There are certain stable arrangements, the electrons falling into well-defined energy levels or shells, defined by quantum conditions. Each energy level can hold no more than two electrons and these must have

opposite spins (see 8 below). This is known as the Pauli exclusion principle.

8. The lowest energy shell, with principal quantum number 1, is the *K* shell. This ground state is called the 1s orbital. Here the electron is distributed with spherical symmetry about the nucleus and has zero angular momentum. However, the electrons can have a spin quantum number which can be described as being either parallel or antiparallel to some specified direction, for example ↑ or ↓. If the K shell contains 1 electron, as in hydrogen, it has a single spin; if it contains two electrons, as in helium, the spins are paired in opposite directions and the K shell is now complete. The atom is thus inert.

9. The next higher energy shell has principal quantum number 2 and is known as the L shell. The lowest level is called the 2s state (again spherically symmetrical) and this can hold 2 electrons of opposite spin. There is a higher energy orbital known as the 2p. This has angular momentum, is dumb-bell in shape, and there are three directions available. Each can hold 2 electrons of paired spins so that altogether this state can hold 6 electrons. If only two or three electrons have to go into the 2p state the energy is lowest if they each have parallel spins (this is part of a more general principle, due to F. Hund, which says that when several levels within a given state are available the lowest energy corresponds to parallel spins): they can do this by each going into the three available orbitals. These electrons can readily play a part in chemical bonding and other processes. A fourth electron will have to pair its spin with one of the existing three. The remaining two unpaired electrons are then readily available for chemical reaction. When the 2p state is filled with its six electrons the atom – neon – is chemically non-reactive (but see below).

10. The next higher level has principal quantum number 3 and is the M shell. This can hold 2 electrons in the 3s orbital, 6 in the 2p and 10 in the 3d. (Again, like the p orbitals, the d orbitals are directed and have angular

Table 1. *Shells and stable sub-groups*

| Shell | Name | Number of electrons (maximum) | Number of electrons in stable sub-groups |
|-------|------|-------------------------------|------------------------------------------|
| 1st | K | 2 | 2 |
| 2nd | L | 8 | 2, 6 |
| 3rd | M | 18 | 2, 6, 10 |
| 4th | N | 32 | 2, 6, 10, 14 |

momentum.) The next higher energy level is the N shell which contains 2 electrons in the 4s, 6 in the 4p, 10 in the 4d and 14 in the 4f states. However, it happens that the lowest state in the N shell (4s) generally has lower energy than the 3d states in the M shell: electrons will therefore go into the 4s state before they start filling the 3d orbitals. The arrangement of shells and stable sub-groups are summarized in Table 1.

As mentioned above if an atom contains a complete shell or sub-group it is relatively stable and may even be inert. In Table 2 we construct the first part of the periodic table based on the shell-model described above.

Table 2. *The beginning of the periodic table of elements*

| Principal quantum numbers | | 1 | 2 | | 3 | | | 4 | | | |
|---|---|---|---|---|---|---|---|---|---|---|---|
| Shell | | K | L | | M | | | N | | | |
| Electronic states | | 1s | 2s | 2p | 3s | 3p | 3d | 4s | 4p | 4d | 4f |
| At. No. | Atom | | | | | | | | | | |
| 1 | H | ↑ | | | | | | | | | |
| 2 | He | ⇅ | inert | | | | | | | | |
| 3 | Li | filled | ↑ | | | | | | | | |
| 4 | Be | „ | ⇅ | | | | | | | | |
| 5 | B | „ | filled | ↑ | | | | | | | |
| 6 | C | „ | „ | ↑, ↑ | | | | | | | |
| 7 | N | „ | „ | ↑, ↑, ↑ | | | | | | | |
| 8 | O | „ | „ | ⇅, ↑, ↑ | | | | | | | |
| 9 | F | „ | „ | ⇅, ⇅, ↑ | | | | | | | |
| 10 | Ne | „ | „ | ⇅, ⇅, ⇅ | inert | | | | | | |
| 11 | Na | „ | „ | filled | ↑ | | | | | | |
| 12 | Mg | „ | „ | „ | ⇅ | | | | | | |
| 13 | Al | „ | „ | „ | filled | ↑ | | | | | |
| 18 | Ar | „ | „ | „ | „ | ⇅, ⇅, ⇅ | inert | | | | |
| 19 | K | „ | „ | „ | „ | filled | empty | ↑ | | | |
| 20 | Ca | „ | „ | „ | „ | „ | empty | ⇅ | | | |
| 21 | Sc | „ | „ | „ | „ | „ | ↑ | ⇅ | | | |
| 29 | Cu | „ | „ | „ | „ | „ | filled | ↑ | | | |
| Maximum occupancy number | | 2 | 2 | 6 | 2 | 6 | 10 | 2 | 6 | 10 | 14 |
| Total | | 2 | 8 | | 18 | | | 32 | | | |

We see that the outermost electrons, especially those with unpaired spins, determine the chemical properties of the atoms. As a simple approximation we may say that the valency reflects the way in which the atom loses, gains or shares electrons in an attempt to form stable shells or sub-shells. When the atom already contains such a configuration (He, Ne, and Ar above) it is already stable and is chemically inert. This needs some qualification since in recent years it has been shown that the larger inert atoms such as xenon can, in fact, be ionized by the presence of a powerful electronegative ion such as the fluoride ion. As a result it is possible to form a stable compound of xenon hexafluoride.

### 1.1.8 *Size of atoms*

The simplest way of forming some quantitative idea of the size of atoms is to measure the distance between atoms in the solid state. The result, of course, depends on the crystal structure. Some typical results are given in Tables 3 and 4.

Table 3. *Atomic diameters for vertical columns in periodic table*

| Atom | $Z$ | Electron shells | | | | | | $d$ (nm) |
|------|-----|---|---|---|---|---|---|----------|
|      |     | K | L | M | N | O | P |          |
| H    | 1   | 1 |   |   |   |   |   | —        |
| Li   | 3   | 2 | 1 |   |   |   |   | 0.3      |
| Na   | 11  | 2 | 8 | 1 |   |   |   | 0.37     |
| K    | 19  | 2 | 8 | 8 | 1 |   |   | 0.46     |
| Rb   | 37  | 2 | 8 | 18| 8 | 1 |   | 0.49     |
| Cs   | 55  | 2 | 8 | 18| 18| 8 | 1 | 0.52     |

In Table 3 the outermost configuration is unchanged. As we fit in additional shells or sub-shells there is an increase in $d$. (The inner orbitals contract because of the increase in nuclear charge.)

In Table 4 the main process is one of filling up an unfilled shell whilst the outer configuration scarcely changes. The increasing nuclear charge causes shrinkage in the atomic diameter.

### 1.1.9 *Size of ions*

Some results are given in Table 5 for monovalent atoms which either lose an electron to become a positive ion, or gain an electron to become a negative ion. It is seen that the effect of removing one electron

Table 4. *Atomic diameters for horizontal columns in periodic table*

| Atom | $Z$ | Electron shells | | | | $d$ (nm) |
| | | K | L | M | N | |
| --- | --- | --- | --- | --- | --- | --- |
| Ca | 20 | 2 | 8 | 8; 0 | 2 | 0.4 |
| Sc | 21 | 2 | 8 | 8; 1 | 2 | 0.35 |
| Ti | 22 | 2 | 8 | 8; 2 | 2 | 0.30 |
| V | 23 | 2 | 8 | 8; 3 | 2 | 0.26 |
| Cr | 24 | 2 | 8 | 8; 5 | 1 | 0.26 |
| Mn | 25 | 2 | 8 | 8; 5 | 2 | 0.26 |
| Fe | 26 | 2 | 8 | 8; 6 | 2 | 0.25 |
| Co | 27 | 2 | 8 | 8; 7 | 2 | 0.25 |
| Ni | 28 | 2 | 8 | 8; 8 | 2 | 0.25 |
| Cu | 29 | 2 | 8 | 8; 10 | 1 | 0.26 |

Note that Ir($Z = 77$) has smaller diameter than Al ($Z = 13$); $d$ (Ir) = 0.27 nm, $d$ (Al) = 0.28 nm.

from the atom is to reduce $d$ by 0.18–0.2 nm; the effect of adding an electron is to increase $d$ by about the same amount.

One obtains a better idea of what is involved by comparing $Na^+$, Ne and $F^-$. These have the same final electron configuration, i.e. are isoelectronic but the nuclear charge is 11, 10, 9 respectively. Consequently one would expect the diameter of $Na^+$ to be smaller than Ne and the diameter of Ne to be smaller than that of $F^-$. This is indeed the case, the values being 0.20, 0.25 and 0.27 nm for $Na^+$, Ne and $F^-$ respectively.

### 1.1.10 *Density of solid elements*

An increase in atomic number usually implies an increase in atomic weight. There is also some increase or decrease in atomic diameter

Table 5. *Effect of ionization on size*
(size obtained from spacing in appropriate compounds)

| Metals | | | Non-metals | | |
| Element | Diameter $d$ (nm) | | Element | Diameter $d$ (nm) | |
| | Atom | Ion | | Atom | Ion |
| --- | --- | --- | --- | --- | --- |
| Li | 0.3 | 0.136 | F | 0.13 | 0.27 |
| Na | 0.37 | 0.196 | Cl | 0.20 | 0.36 |
| K | 0.46 | 0.296 | Br | 0.23 | 0.39 |
| Rb | 0.49 | 0.296 | I | 0.27 | 0.44 |
| Cs | 0.52 | 0.330 | | | |

(see Tables 3 and 4) but it is not as large a change as the change in atomic weight. Consequently although there is generally an increase in density of the solid element as the atomic number increases it is not necessarily a monotonic relation.

It is interesting to note that if we compare the atomic weight of hydrogen with that of uranium $(1:238)$ it is not very different from the ratio of their densities $(70:18\,400\ \mathrm{kg\,m}^{-3})$.

## 1.2 The forces between atoms and molecules

### 1.2.1 *Forces due to the ionic bond*

The simplest forces between atoms are those which arise as a result of electron transfer. A simple example is that of, say, sodium fluoride. The sodium atom has a nuclear charge of $+11$, with 2 electrons in the K shell, 8 in the L shell and 1 in the M shell. The fluorine atom has a nuclear charge of 9 with 2 electrons in the K shell, and 7 in the L shell. The outermost electron in the sodium atom may transfer readily to the fluorine atom; both atoms then have a complete shell but the sodium now has a net charge of $+1$ and the fluorine a net charge of $-1$. These ions therefore attract one another by direct coulombic interaction. The force between them is strong – it varies as $x^{-2}$, where $x$ is the distance between the ions, and it acts in the direction of the line joining the ions. Furthermore it is unsaturated – one positive ion can attract several negative ions around it and the force exerted by the positive ion on each negative ion is not affected by the presence of other negative ions. Of course the negative ions will also repel one another.

### 1.2.2 *Forces due to the covalent bond*

These are sometimes referred to as valency forces, for they are the type of force which accounts for the binding together of atoms in those molecules where the primary attraction is not ionic. For example the

Figure 3. Electron concentration in a pair of hydrogen atoms; (*a*) unstable, the molecule dissociates into two atoms, (*b*) stable, the electron concentration between the nuclei serves to bond the nuclei together as the $H_2$ molecule.

force binding two hydrogen atoms together to form a hydrogen molecule is of this nature. Covalent bonding always involves the sharing of electrons. One way of describing this is to say that in the hydrogen molecule the single electron of each atom interacts with both nuclei, so that each atom regards itself as 'possessing' two electrons which provide a complete, stable 'molecular' K shell. However, the binding cannot be described in classical terms. It arises from the fact that the electrons $e_1$ and $e_2$ are indistinguishable, so that if $e_1$ and $e_2$ are interchanged this does not involve a new configuration, but leads to an 'exchange energy' which provides the force binding the atoms together. Descriptively we may say that there is an appreciable concentration of electrons (negative charge) between the two nuclei (positive charge), and this binds the atoms together. This is indicated schematically in figure 3.

Covalent forces are strong, they fall off rapidly with separation; they act in specified directions (valency bonds) and they are saturated.

### 1.2.3 *Van der Waals forces*

Van der Waals forces exist between all atoms and molecules whatever other forces may also be involved. We may divide van der Waals forces into 3 groups.

(*a*) *Dipole–dipole forces.* If we consider a substance such as HCl, where the bonding is largely ionic, the individual molecule will consist of a positive and a negative charge separated by a small distance of the order of 0.1 nm, so that it constitutes a minute electric dipole. This will interact with a neighbouring HCl molecule and produce a net attractive force. The force (see (*c*) below) will be proportional to $x^{-7}$, where $x$ is the distance between the molecules.

(*b*) *Dipole-induced dipole forces.* Consider the interaction between say an HCl molecule and a neon atom. The latter has no natural dipole. However, the HCl dipole will induce a dipole in the neon atom (see (*c*) below) and there will then be an attractive force between them again proportional to $x^{-7}$.

(*c*) *Dispersion forces.* The most 'mysterious' type of force is that which exists between two non-polar atoms such as neon. We know that these atoms must attract one another, for otherwise we could never obtain liquid or solid neon simply by cooling the material and so reducing the thermal energy of the atoms below some critical level. The explanation is as follows. Although the neon atom has a symmetrical distribution of electrons around the nucleus so that the atom has no dipole, this statement is true only as a time average. At any instant there may be some asymmetry in the distribution of the electrons around the nucleus. The neon atom at this instant, therefore, behaves as an electric dipole. This produces a field which distorts the electron distribution in a neighbouring atom (polarizes it) so that the neighbouring atom itself acquires a dipole. The two dipoles will attract. This is the basis of the 'dispersion' van der Waals force.

We can easily derive the law of force in the following way. Suppose that at a given moment the first atom has a dipole moment $\mu$ (figure 4). At a distance $x$ along the axis of the dipole the electrostatic field $E$ is given by

$$E = \frac{2\mu}{x^3}\frac{1}{4\pi\varepsilon_0}, \tag{1.10}$$

where $\varepsilon_0$ is the permittivity of space. (This equation is only valid if $x$ is reasonably large compared with the length of the dipole in figure 4. For closer approach, a more complicated equation would be needed.) If there is another dipole at this point it will be polarized and acquire a dipole $\mu'$. In order to describe this we invoke a property discussed in Chapter 12: that the dipole produced by a field is approximately proportional to the strength of the field. We write

$$\mu' = \alpha E, \tag{1.11}$$

where $\alpha$ is termed the atomic polarizability of the atom.

Now a dipole $\mu'$ in a field $V$ has a potential energy $V$ given by

$$V = -\mu' E$$
$$= -\alpha E^2 = \frac{-4\alpha\mu^2}{x^6}\left(\frac{1}{4\pi\varepsilon_0}\right)^2. \tag{1.12}$$

This implies a force between the atoms of magnitude

$$F = -\frac{\partial V}{\partial x} = \frac{\text{constant}}{x^7}. \tag{1.13}$$

We see that the force is always attractive and falls off as $x^{-7}$.

We may take this two stages further. For simplicity let us consider the attraction between two hydrogen *atoms*, where only one electron can be

Figure 4. A dipole produces an electric field at a distance $x$ along its axis of amount $E = (2\mu/x^3)(1/4\pi\varepsilon_0)$. This field can polarize another molecule such that mutual attraction occurs.

involved in each atom. If the effective radius of the atomic orbital is $r$, the maximum value of the instantaneous dipole is $re$. Substituting in equation (1.12) we have

$$V = -\frac{4\alpha r^2 e^2}{x^6}\left(\frac{1}{4\pi\varepsilon_0}\right)^2. \tag{1.14}$$

We now make use of a derivation given in Chapter 12. For a simple atom, or for a small molecule which is virtually spherically symmetrical, the polarizability is given by

$$\alpha = r^3(4\pi\varepsilon_0), \tag{1.15}$$

where $r$ is the radius of the atom or molecule. For reasons which will soon become apparent we select from this relation the value of $r^2$

$$r^2 = \frac{\alpha}{r}\frac{1}{4\pi\varepsilon_0}. \tag{1.16}$$

Insertion in equation (1.14) gives

$$V = \frac{-4\alpha}{x^6}\frac{\alpha}{(4\pi\varepsilon_0)^2}\frac{e^2}{r(4\pi\varepsilon_0)}. \tag{1.17}$$

By comparison with equation (1.8) we see that the last term on the right is the quantized energy level $h\nu$ of the electron when its frequency is $\nu$. We are left with

$$V = -\frac{4\alpha^2}{x^6}h\nu\left(\frac{1}{4\pi\varepsilon_0}\right)^2. \tag{1.18}$$

This derivation is, of course, rather crude since it assumes that the dipoles are always oriented end-on and we have ignored the fact that the second atom must itself influence the polarization of the first. However, it contains all the essential features of the van der Waals interaction: the polarizability $\alpha$, the separation $x$ and the frequency $\nu$ of the polarized orbital. The rigorous derivation by F. London, in 1930, considered the perturbation produced in the solution of the Schrödinger equation. His result is

$$V = -\frac{3}{4}\frac{\alpha^2}{x^6}h\nu\left(\frac{1}{4\pi\varepsilon_0}\right)^2 \tag{1.19}$$

and is exactly the same as our simple derivation except for the numerical factor. As we shall see in Chapter 12 the frequency $\nu$ is the same as that which accounts for the dependence of refractive index on the frequency of the incident light, i.e. it is the frequency involved in optical dispersion. For this reason those van der Waals forces which arise by mutual polarization of the atoms or molecules are referred to as dispersion forces.

Van der Waals forces are much weaker than ionic and covalent forces. They are central and, like ionic forces, they are unsaturated: also to a first approximation they are additive (see below).

### 1.2.4 *Retardation effects in van der Waals forces*

The derivation of the London relation is valid only if the atoms are less than a few hundred angstroms, i.e. a few tens of nm, apart for the following reason. For large separations the electric field from one atom takes an appreciable time to reach its neighbour: by the time it has done so and polarized it, it itself will have acquired a different electron configuration. There is a lag in the interaction and this leads to what is known as retarded van der Waals forces. In different words we may say that when the atoms are close together the electron fluctuations between neighbours are in close correlation: when they are far apart this correlation gets poorer and poorer, and the potential now falls off as $1/x^7$. We can form some estimate of the distance at which non-retarded forces go over to retarded forces by noting that an orbiting Bohr electron takes about $3 \times 10^{-16}$ s to rotate about the nucleus so that an electric field, travelling at the velocity of light $(3 \times 10^8 \text{ m s}^{-1})$ will cover a distance of about $10^{-7}$ m or 100 nm in this time. Thus when atoms are more than this distance apart we should expect the correlation to be lost and the interaction to be almost completely retarded. Experiments described below show that, in fact, the transition from non-retarded to retarded forces occurs at a separation of the order of 10 to 50 nm.

There is no simple derivation of retarded van der Waals forces but the following argument provides some insight into its origin. The first point to emphasize is that retarded van der Waals forces are not basically a new type of interaction. All that really happens is that the atoms select from the fluctuations in the electron distribution those which occur at a low enough frequency to give good correlation between neighbouring atoms. We replace $\nu$ in equation (1.19) by $c/\lambda$, where $c$ is the velocity of light and $\lambda$ the wavelength of the fluctuation. We now specify that if $\lambda$ is of the same order as the separation $x$ (or greater) fluctuations of this wavelength in one atom will be in good correlation with fluctuations in another atom at distance $x$.† All shorter wavelengths (i.e. higher frequencies) will give

---

† Sophisticated theory shows that the loss in correlation occurs at distances greater than $\lambda/2\pi$ rather than $\lambda$: but this does not change the type of argument given here.

such poor correlation that their contributions to the interaction energy can be neglected. We thus convert equation (1.19) into

$$U = -\frac{3}{4}\frac{\alpha^2}{x^6}\frac{hc}{\lambda}\left(\frac{1}{4\pi\varepsilon_0}\right)^2,$$

where $\lambda$ is of order $x$. This gives as a final result

$$U \approx -\frac{3}{4}\frac{\alpha^2}{x^7}hc\left(\frac{1}{4\pi\varepsilon_0}\right)^2, \tag{1.20}$$

so that the attractive force now falls off as $1/x^8$. Equation (1.20) is very close to the correct answer obtained by extremely difficult theory (Casimir gives $23/8\pi^2$ instead of $\frac{3}{4}$ in equation (1.20)).

van der Waals forces exert their influence in the behaviour of individual atoms or molecules in the gaseous state (see Chapter 5). The magnitude of these forces can then be deduced by comparing the behaviour of a real gas where these forces operate with that of an ideal gas where intermolecular forces are negligible. A more direct way of determining the magnitude of these forces is by measuring the forces between macroscopic bodies. We assume that the forces are roughly additive and carry out a pair-wise addition for each atom in one body with every atom in the other body. Additivity is not strictly true but is a reasonably good first approximation. For a sphere of radius $R$ at a distance $x$ of *nearest* distance to a flat surface (or for crossed cylinders each of radius $R$) the force is

$$F_{\text{non-retarded}} = AR/6x^2, \tag{1.21a}$$

$$F_{\text{retarded}} = 2\pi BR/3x^3, \tag{1.21b}$$

where $A$ and $B$ are known as the Hamaker constants for non-retarded and retarded forces respectively. Experimental results for crossed cylinders of mica are shown in figure 5. It is seen that the power law of $x$ changes from 2 to 3 over the range of 10 to about 50 nm. The results also show that for mica the constant $A$ has a value of about $1.3 \times 10^{-19}$ J and $B$ a value of about $10^{-28}$ J m. Retarded forces are not of great importance since they are small and play little part in most physical processes. By contrast the non-retarded forces can be relatively large, particularly if the bodies come into atomic contact. Values of $A$ range from about $10^{-19}$ J for inorganic solids to about $3 \times 10^{-19}$ J for metals. If the bodies are immersed in a liquid the value of $A$ is less, often by about an order of magnitude.

For extended parallel surfaces the forces are naturally proportional to the area of the surfaces. The force $f$ per unit area has the value

$$f_{\text{non-retarded}} = \frac{A}{6\pi x^3} \qquad (1.22a)$$

and

$$f_{\text{retarded}} = \frac{B}{x^4}. \qquad (1.22b)$$

Consider two smooth polished surfaces placed in contact and kept apart by a few asperities or a few dust particles so that the separation is of the order 1 $\mu$m. Then, assuming retarded forces and a value of $B \approx 10^{-28}$ J m, the force $f$ is of order $10^{-4}$ N m$^{-2}$, i.e. an attractive force per square metre about equal to the weight of a fly. If, however, the separation were reduced to atomic dimensions, say 0.2 nm, and if the surfaces were molecularly smooth, the non-retarded force would be of order $10^9$ N m$^{-2}$, i.e. 100 kg mm$^{-2}$: this approaches the strength of real solids. These simple calculations show that the range of action of surface forces is

Figure 5. Variation of power-law index $n$ of the van der Waals law of force between crossed mica cylinders with distance of separation $x$. Below 10 nm the forces are non-retarded and the law of force is $F = AR/6x^n$ with $n = 2$, where $R$ is the radius of the cylinders. Above 50 nm the forces are retarded and the law of force is $F = 2\pi BR/3x^n$, where $n = 3$.

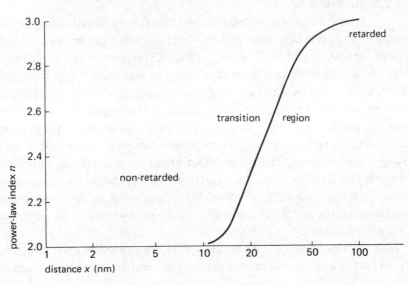

small: in theory these forces extend to infinity but their magnitude is appreciable only for separation of the order of a few atomic diameters.

The consequences of this may at once be seen. Even in the absence of ionic or valency forces, surfaces will stick together strongly if they can come into close contact (within a few angstroms) of one another. Again gases will generally be adsorbed strongly for the first monolayer but the attraction falls off so rapidly with distance that second and third layers are far less likely. Roughly speaking the formation of adsorbed layers depends on whether the potential energy drops by a greater amount than the mean thermal energy. For a molecule, the latter is of the order of $kT$, where $k$ is the Boltzmann constant and $T$ the absolute temperature (see Chapter 3). This is not the whole story since, of course, the latent heat of condensation is also a factor encouraging further adsorption, whilst the entropy decrease associated with an 'ordered' adsorbed film (see Chapter 2) tends to oppose adsorption. The net result is that polymolecular adsorption is unlikely unless the gas or vapour is very near saturation.

With large particles, such as those in colloidal systems or Brownian suspensions, the position is different. The thermal energy of the individual particle is still given by $kT$ but the van der Waals energy must be integrated for the millions of atoms in the particles. In such systems the van der Waals forces can in fact play a basic part in determining whether particles will stick together or not. This will be discussed in greater detail at the end of Chapter 11.

### 1.2.5 *Repulsion forces*

If atoms were subjected only to attractive forces all atoms would coalesce. Thus it is clear that for very short distances of separation some type of repulsive force must operate. In ionic systems the attractive force arises from the electrostatic charge and the repulsive force from the electron shells, which act as a sort of tough elastic sphere resisting further compression. The repulsion may be described as arising from two effects. First, the penetration of one electron shell by the other, which means that the nuclear charges are no longer completely screened and therefore tend to repel one another. The other effect arises from the Pauli exclusion principle which states that two electrons of the same energy cannot occupy the same element of space. For them to be in the same space (overlapping) the energy of one must be increased – this is equivalent to a force of repulsion.

With covalent and van der Waals forces the repulsion and attraction are all part of a single mechanism and it is not correct to consider them as arising from separate mechanisms; however, for the purpose of simple calculation it is very convenient to do so. We therefore describe the long-range attraction by a term of the form

$$F = \frac{A}{x^m},\qquad(1.23)$$

where $A$ and $m$ are suitable constants, and the repulsion by a relation of the form

$$F = \frac{B}{x^n},\qquad(1.24)$$

where $B$ is a constant and $n$ has a value of the order 8–10. In fact the power-law relation is not very good and there are theoretical reasons for preferring a relation for the repulsive force of the form $Be^{-x/s}$ where $B$ and $s$ are suitable constants. In the following calculation we shall for simplicity retain the power law for both attraction and repulsion forces.

The resultant force is then

$$F = \frac{A}{x^m} - \frac{B}{x^n}\qquad(1.25)$$

and this is plotted in figure 6. The equilibrium separation $x_0$ occurs when $F = 0$. This gives $B = Ax_0^{n-m}$.

Figure 6. (a) Force and (b) potential energy curves for two atoms (or molecules) as a function of separation. There is a long-range attractive force and a short-range repulsive force which operates only when the molecules come close together. The equilibrium separation occurs when the net force is zero (at $A$) or when the potential energy is a minimum.

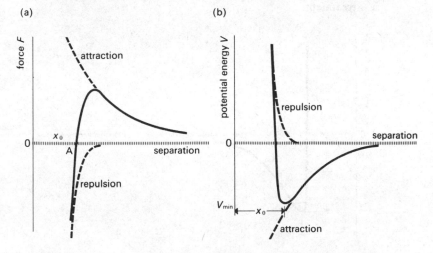

Hence equation (1.25) becomes

$$F = A\left(\frac{1}{x^m} - \frac{x_0^{n-m}}{x^n}\right).$$ (1.26)

We may at once note that for small displacements from the equilibrium position the restoring force is

$$dF = A\left(\frac{n-m}{x_0^{m+1}}\right) dx.$$ (1.27)

Since the force is proportional to the displacement, the motion of the displaced particle will be simple harmonic. This conclusion is of course not restricted to power-law relations. Any relation of the form $F = f(x)$ will lead to a similar conclusion.

### 1.2.6 *Potential energy*

The potential energy $V$ involved in bringing one atom from infinity to a distance $x$ from another atom is given by

$$V = \int (\text{external force}) \, dx = -\int (\text{internal force}) \, dx = \int F \, dx,$$ (1.28)

since $F$ acts in the opposite direction to $dx$.

Figure 7. Stylized figure of potential energy curve of one molecule in relation to a single neighbour emphasizing the depth of the potential energy trough $\Delta\varepsilon$, the extremely steep repulsive part of the curve, and the effective diameter $\sigma$ of the molecule. The zero-point energy, $\frac{1}{2}h\nu_0$, where $\nu_0$ is the fundamental frequency of the pair, is very small compared with $\Delta\varepsilon$ and will be neglected in the rest of this treatment.

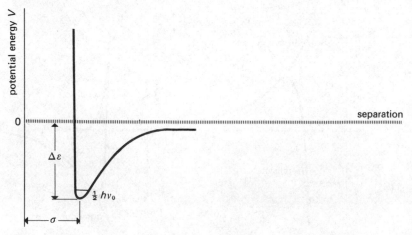

From equation (1.26) this gives

$$V = A\left(-\frac{1}{m-1}\frac{1}{x^{m-1}} + \frac{1}{n-1}\frac{x_0^{n-m}}{x^{n-1}}\right).$$
$$\underbrace{\phantom{-\frac{1}{m-1}\frac{1}{x^{m-1}}}}_{\text{attractive term}}\quad\underbrace{\phantom{\frac{1}{n-1}\frac{x_0^{n-m}}{x^{n-1}}}}_{\text{repulsive term}}$$

A potential energy equation of this form was first proposed by Mie in 1903. Note that the attractive term gives a negative potential energy, the repulsive term a positive potential energy (see figure 7).

At the equilibrium position, where $x = x_0$, the potential energy has a minimum value

$$V_{min} = -A\frac{1}{x_0^{m-1}}\left(\frac{1}{m-1} - \frac{1}{n-1}\right).$$

The interatomic forces are summarized in Table 6.

Table 6. *Interatomic forces*

| Type | Nature | $A$ | $m$ | Energy to separate 2 atoms or ions |
|------|--------|-----|-----|-------------------------------------|
| Ionic | Central unsaturated | Very large | 2 | $Na^+ + F^-$; $10^{-18}$ J |
| Covalent | Directed saturated | Very large | [a] | H–H; $7 \times 10^{-19}$ J |
| van der Waals | Central unsaturated | Small to very small | 7 | Ne: Ne; $5 \times 10^{-22}$ J |

[a] The attractive force between two hydrogen atoms in the hydrogen molecule is roughly proportional to $1/x^3$. This only applies for small displacements; it cannot be applied over large displacements nor can it be considered as being valid for covalent forces in general.

### 1.2.7 Intermolecular forces

Intermolecular forces are of the same nature as those occurring between atoms. For ionic crystals such as rock-salt one cannot distinguish individual molecules of NaCl, as the crystal is an array of positive ($Na^+$) and negative ($Cl^-$) ions held together by coulombic forces. In covalent solids such as diamond the individual carbon atoms are joined, by valency bonds, to form a strongly-linked crystal. In a solid such as paraffin wax, or solid hydrogen, the molecules are held together by weak van der Waals forces. However, in a polymer these may be augmented by the attraction between polar groups or by chemical bonds. There is an additional type of bond which occurs only in the aggregate; this is the metallic bond. Individual metal atoms exist as such in the vapour phase, where a small number of diatomic molecules may occur, but there little tendency for the outermost valency electrons to join forces and produce polyatomic molecules in the vapour phase. There is, however, strong bonding in the condensed solid state. The atoms become ionized and exist as positive ions in a sea of free valency electrons; they are held together by the strong attraction between the ions and the electrons. In effect, energy has to be provided to ionize the metal atoms, but this is more than compensated for by the binding energy between the ions and the electron sea.

The simplest measure of the intermolecular forces is the heat of sublimation, which is the amount of energy needed to separate all the molecules from all their neighbours. This is highest for ionic and covalent solids, high for metals and least for van der Waals solids. We may make this more quantitative in the following way. If the heat of sublimation $E_s$ is given in calories per mole, we multiply it by 4.2 to convert to Joules and divide by $6 \times 10^{23}$ to obtain the energy in Joules per single molecule. This conversion factor is $7 \times 10^{-24}$. The result is the energy $E_s$ required to break bonds between one molecule and all its neighbours. If, therefore, the coordination number (number of nearest neighbours) is $n$ and only nearest-neighbour interaction is involved, we may write for the binding energy $\varepsilon$ between one molecule and one neighbour

$$E_s = \tfrac{1}{2} n \varepsilon.$$

The $\tfrac{1}{2}$ is introduced since otherwise every bond would be counted twice. We then have

$$E_s \times 7 \times 10^{-24} = \frac{n}{2} \varepsilon \quad \text{or} \quad \varepsilon = \frac{14 \times 10^{-24}}{n} E_s.$$

As an example we quote the simplest type of solid–solid neon, which forms a close-packed face-centred cubic (f.c.c.) structure of the van der Waals type. A sketch illustrating this structure is given in Chapter 7, figure 54. The latent heat of sublimation is found by experiment to be approximately 450 calories per mole. The number of nearest neighbours, $n$, is 12.

Hence

$$\varepsilon \approx 5 \times 10^{-22} \text{ J}.$$

This agrees well with the theoretical interaction energy between one neon atom and its neighbour.

Another example, which is less direct, is worth quoting because it provides a check on our deduced value of $\varepsilon$ by a completely different method. In sodium fluoride the energy to convert 1 mole of NaF to $Na^+$ and $F^-$ free ions is 213 000 cal. This is approximately $15 \times 10^{-19}$ J per molecule of NaF or $7.5 \times 10^{-19}$ J per ion of $Na^+$ or $F^-$. In the crystal lattice each ion is surrounded by 6 ions of opposite sign (i.e. $n = 6$). The energy per unit bond (it is of course negative) is therefore

$$\varepsilon = \frac{7.5 \times 10^{-19}}{3} = 2.5 \times 10^{-19} \text{ J}.$$

We may now consider the ionic interaction between a $Na^+$ and a $F^-$ as they are separated from their equilibrium position in the lattice (separation $x_0 = 0.231$ *nm*) to infinity. Ignoring the repulsion term, we have

$$\varepsilon = -\frac{1}{4\pi\varepsilon_0} \int_{x_0}^{\infty} \frac{e^2}{x^2} \, dx = -\frac{1}{4\pi\varepsilon_0} \frac{e^2}{x_0} = -10^{-18} \text{ J}.$$

This assumes only nearest-neighbour interaction. There is, in fact, a very large interaction between the next nearest neighbours. For example the potential energy due to the attraction between one positive ion and the negative ions surrounding it is greatly reduced by the coulombic repulsion between the positive ion and the nearest positive ions. These are only a little further away than the nearest negative charges and as coulombic potentials fall off only as $1/x$, the resultant potential energy is greatly reduced. Detailed calculations such as those indicated in Chapter 9 show that the energy is reduced from $-(4\pi\varepsilon_0)^{-1}e^2/x_0$ to 0.2905 times

Table 7. *Intermolecular forces*

| Type | Nature | Example | Mode of separation | Energy per bond $J \times 10^{21}$ |
|------|--------|---------|--------------------|-----------------------------------|
| Ionic | Identical with interatomic forces | NaF crystal | Break up into $Na^+$ and $F^-$ ions | 300 |
| Covalent | Giant molecule held by covalent forces | Diamond | Break up into C atoms | 600 |
| van der Waals | As for atoms | Solid methane $CH_4$ | Break up into $CH_4$ gas | 3 |
| Metal | Metal ions in sea of free valency electrons | K | Vaporize to metal ions + electrons | 500 |

this value. This gives a value for $\varepsilon$ of about $2.9 \times 10^{-19}$ J which agrees well with the value quoted at the top of p. 27.

It should be emphasized that the calculation of intermolecular forces from the experimentally determined heat of sublimation can only be applied in a direct way to van der Waals solids. In this case the sublimed material consists of the same chemical units as those existing in the solid. With ionic solids such as NaF sublimation breaks up the solid into a vapour of NaF molecules. Thus the energy involved in separating all the $Na^+$ and $F^-$ ions to infinity is greater than the heat of sublimation by the dissociation energy of NaF. It is this quantity which was used in the calculation at the top of p. 27. (We ignore the kinetic energy of the vapour molecules or ions.) Again when a metal is sublimed the vapour consists primarily of free atoms: consequently the energy involved in separating the metal ions and the binding electrons is greater than the sublimation energy by, approximately, the ionization potential of the metal. Some of these issues are discussed in greater detail in Chapter 7, where it is shown that the elastic constants of solids can be calculated in terms of intermolecular forces. At this stage we merely summarize the main features of intermolecular forces in Table 7.

### 1.2.8. *Comparison of electrical and gravitational attraction*

Consider the interaction between, say, two helium atoms, where only the weakest electrical attraction exists as a result of the van der Waals mechanism. If the atoms are packed together as close as possible (say in liquid state) the separation $x_0$ is about 0.2 nm. The potential energy $V_W$ due to van der Waals for this separation turns out to be about $-10^{-22}$ J. Now consider the potential energy $V_g$ due to gravity.

$$V_g = \int_{x_0}^{\infty} G\frac{m_1 m_2}{x^2}\, dx = -G\frac{m_1^2}{x_0} = -\frac{6.7 \times 10^{-11}(0.7 \times 10^{-26})^2}{2 \times 10^{-10}},$$

i.e.

$$V_g = -10^{-53} \text{ J},$$

whereas

$$V_W = -10^{-22} \text{ J}.$$

We see that even when only van der Waals forces are involved the gravitational energy is trivial compared with the electrical.

### 1.2.9 *The simplified potential energy curve*

In this section we shall treat molecules as though they are hard elastic balls of diameter $\sigma$ which attract each other according to some suitable law of force, and then experience a very powerful repulsion force for separations less than $\sigma$.

The resulting potential energy curve is then as shown in figure 7 and the two most important factors are $\sigma$ and the depth $\Delta\varepsilon$ of the potential energy trough. Some typical values are given in Table 8.

Table 8

| Molecule | $\sigma$ (nm) | $\Delta\varepsilon$ (J) (approx.) |
|---|---|---|
| He | 0.22 | $1 \times 10^{-22}$ |
| $H_2$ | 0.27 | $4 \times 10^{-22}$ |
| Ar | 0.32 | $16.5 \times 10^{-22}$ |
| $N_2$ | 0.37 | $13 \times 10^{-22}$ |
| $CO_2$ | 0.45 | $40 \times 10^{-22}$ |

In later chapters we shall be able to explain, in terms of such a potential energy curve, a large number of phenomena. These include critical temperature of a gas, surface tension and viscosity of a liquid, heat of sublimation of a solid, elastic constants and the thermal expansion of a solid.

# 2

# Temperature, heat and the laws of thermodynamics

In this chapter we consider briefly some of the basic concepts of heat and temperature. The treatment is simple and restricted to those aspects which will be of use in later parts of the book.

## 2.1 Temperature

### 2.1.1 *The concept of temperature*

The concept of temperature originally arose from a sensory feeling of hot or cold. Probably all physical concepts arise in a similar way; the next step is to turn the idea into something more general and, in particular, more objective. One such approach is as follows. It is found that a given mass of gas is completely specified by its volume $V$ and its pressure $P$. Suppose we start with 1 kg of a gas and subject it to any changes we wish, whether by compressing it, cooling it, allowing it to expand, heating it; we then take another 1 kg specimen of the same gas and carry out a completely different series of operations. If we end up with the same values of $V$ and $P$ for the two samples, the final states will be found identical in every way – colour, warmth, viscosity, and every objective and subjective test to which we can subject the sample.

If we end up with different $P$, $V$ values there is something different about the samples and if they are placed in contact changes take place until they reach identical $P$, $V$ values (see section 2.1.3 below).

The attribute which produces this difference is the temperature, and the changes brought about when the specimens are placed in contact result in their eventually reaching the same temperature. For gases and fluids there is a unique function of $(P, V)$ which determines their temperature $\theta$ if the mass is specified: $f(P, V) = \theta$. For solids one needs more parameters.

### 2.1.2 *Temperature scales*

There are, of course, sophisticated means of deriving an absolute thermodynamic scale of temperature (see section 2.3.6 below). In practice one can use the $P-V$ properties of a gas. One may also use less direct standards such as the expansion of a metal rod, the expansion of a liquid column (a conventional thermometer), or the e.m.f. produced by a thermocouple.

### 2.1.3 *Zeroth law of thermodynamics*

In the development of thermodynamics the first two laws were developed and so named, before it was realized that a preliminary law was required to satisfy a rigorous formulation of thermodynamic laws. This was therefore called the zeroth law. (It has nothing to do with Professor Zero.) The law simply states that if two bodies A and B are individually in thermal equilibrium with a body C, then A and B are in thermal equilibrium with one another.

If C is a thermometer this is clearly a matter of common experience.

## 2.2 **Heat**

### 2.2.1 *The concept of heat*

Suppose we have a body of mass $M_1$ at temperature $\theta_1$, which is placed in thermal contact with another body of mass $M_2$ and temperature $\theta_2$, where $\theta_2$ is greater than $\theta_1$. Equilibrium will be reached at a common temperature $\theta$, $\theta_2 > \theta > \theta_1$. We say that heat has flowed from $\theta_2$ to $\theta_1$.

Without specifying what heat is, we can define its units. In the earlier work the heat required to raise the temperature of 1 gram of water at 15 °C by 1 K was defined as the calorie. If we assume that in any experiment involving the flow of heat from one body to another heat is conserved we may define the specific heat capacity of a given substance in terms of the specific heat capacity of, say, water. The specific heat capacity $s$ would then be defined as the number of calories required to raise the temperature of 1 gram of the substance by 1 K.

We can now carry out experiments with different masses of different materials at various temperatures using only a centigrade scale thermometer. Common experience then shows that all the results are consistent with our original assumption that heat is always exactly conserved. We have to qualify this by adding that this is true only if any volume changes which occur do a negligible amount of work.

So far we have said nothing about the nature of heat, but this does not affect the validity of the calorimetric conservation principle which we have just described. The next step is to consider briefly J. P. Joule's classical experiments which show that heat is, in fact, a form of energy.

### 2.2.2 *Joule's experiments*

J. P. Joule (1818–89) began his famous series of experiments on the relation between heat and energy in 1840 and summarized his major conclusions in a massive *mémoire* published in the *Philosophical Transactions* in 1850. His first experiments (1843) involved the combined effects of mechanical and electrical energy, while his second experiments (1845) were concerned with the heat liberated when a gas is compressed. His more famous experiment in which he directly compared the mechanical work expended in stirring water with the heat evolved did not appear until later in 1845, and a greatly improved version was described by him in 1878. The principle was to stir water in a copper vessel with paddles and to calculate the mechanical work done $W$. Using a very sensitive thermometer he observed the temperature rise $\Delta t$ of the water, paddles and vessel, and estimated heat losses by conduction, radiation and (if relevant in the mechanical arrangements of the driving system) the losses due to friction in the bearings. From subsidiary calorimetric experiments he was able to determine the specific heat capacities of the vessel and the paddles in terms of the unit heat capacity of water. In this way he could calculate the total amount of heat $H$ in a purely calorimetric experiment required to produce the same temperature rise as in the paddle-stirring experiment, viz:

$$H = \Delta t[\text{mass of water} + m_1 s_1 \text{ (vessel)} + m_2 s_2 \text{ (paddles)}] + \text{losses.}$$
(2.1)

He then found that $H$ was closely proportional to the mechanical work $W$.

We may summarize some of his other experiments,

(a) He repeated the stirring experiments with mercury as the liquid.

(b) He rubbed iron rings together against their interfacial friction.

(c) He passed an electric current through a resistance wire immersed in a liquid.

(d) He compressed air in an insulated cylinder.

In all cases he found that the observed temperature rises could be achieved simply by adding heat of an amount proportional to the mechanical or electrical work done. Further, the factor of proportionality was a constant

$$W = JH,$$
(2.2)

where $J$ is Joule's constant or the mechanical equivalent of heat and has the value $J = 4.187 \text{ J cal}^{-1}$.

Joule's constant is of immense importance in all mechanical–thermal operations, but in effect this reduces to translating specific heat capacity data into energy data. It is clearly not a fundamental unit of physics. Indeed if calorimetric studies (involving pounds or grams of water and degrees centigrade or fahrenheit) had not been so attractive as a convenient means of describing thermal energy, it is possible that our basic unit would have been the Joule; we should then have had a calorimetric heat-unit defined as the heat required to raise the temperature of one gram of water by 0.238 K. Indeed books of tables now quote specific heat capacities in Joules per g or per kg per degree C.

### 2.3 The laws of thermodynamics
### 2.3.1 *Internal energy and the first law of thermodynamics*

If we carry out an experiment in a thermally isolated calorimeter, such that no heat can flow into or out of the system, we can change the state of the system by performing mechanical or electrical work on it. Joule's experiments show that for the same amount of work we shall always arrive at the same final state. We may now define the difference in internal energy $U$ of the system as the work done on the system in an adiabatic calorimeter.

Since we can change from state 1 to state 2 by an infinite number of combinations of heat and work, if we wish to define internal energy as a function of state only, we must include both work and heat in our definition of energy. The following example illustrates this in a more concrete way.

Suppose we take a mass of water and perform mechanical and/or electrical work on it; provided we do the same amount of work $W$, we produce the same change in temperature. Alternatively we can apply a quantity of heat $H = W/J$ and produce the same change. If we start with water at 0 °C and end with water at 100 °C, then the water ends with the same energy state whatever the path adopted. For example we could add heat to raise the temperature to 30 °C and then stir paddles to supply the remaining 70 °C rise. Clearly an infinite number of paths is available. Yet the change in internal energy $U_{100} - U_0$ (ignoring any work done by the liquid on expanding) must be determined only by the initial and final temperature, not by the path. If this were not true we should be able to devise heating and cooling cycles whereby we could raise the temperature to 100 °C by one path and cool it back to 0 °C by another, but still obtain energy from the system. This is contrary to Joule's experiments.

We are left with the conclusion that a change in $U$ is a function only of the initial and final states. If we do external work $\Delta W$ and add heat $\Delta Q$ we have

$$\Delta U = \Delta W + \Delta Q, \tag{2.3}$$

where $\Delta U$ is uniquely defined, but $\Delta W$ and $\Delta Q$ can have any values provided their sum is equal to $\Delta U$.

For a specified change in state

$$W_A + Q_A \text{ over path A} = W_B + Q_B \text{ over path B.} \tag{2.4}$$

Thus work plus heat are conserved. We may regard this as an extension of the conservation of energy which is well established in *mechanics* and conclude that:

energy is conserved if we include heat as one of the forms of energy.

This is one way of formulating the first law of thermodynamics.

In the simple example given above we have considered $U$ to consist of heat and mechanical or electrical energy. A little consideration shows that the internal energy can embrace all forms of energy, thermal, mechanical, electrical, gravitational, electromagnetic, etc. It may again be shown by similar arguments that, however many types of energy are involved, any change in the internal energy does not depend on the path; it is a function solely of the initial and final states.

### 2.3.2 *Internal energy of fluids*

For a fluid, $U$ is found to be a unique function of only two variables; either $(P, V)$, $(P, T)$ or $(V, T)$, where $P$ is the pressure, $V$ the volume and $T$ the temperature. Since $U$ is independent of path, changes in it can always be expressed in terms of an exact differential. This also means that, in principle, we can integrate $dU$ between initial and final states and always end up with the same answer. Thus we can always write for $U(T, V)$

$$dU = \left(\frac{\partial U}{\partial T}\right)_V dT + \left(\frac{\partial U}{\partial V}\right)_T dV, \tag{2.5a}$$

for $U(P, V)$

$$dU = \left(\frac{\partial U}{\partial P}\right)_V dP + \left(\frac{\partial U}{\partial V}\right)_P dV. \tag{2.5b}$$

In our example of raising the temperature of water from 0 °C to 100 °C we concentrated on the variation of $U$ with temperature; $(\partial U/\partial T)_V$ in equation (2.5a). That is why we made the proviso that for this to be strictly true any changes in volume would have to be ignored, i.e. $(\partial U/\partial V)_T = 0$.

### 2.3.3 *Reversible changes in a gas*

If we allow a gas to expand, it does work against its surroundings. The work done depends on the pressure of the gas, the volume change and the pressure outside the gas against which it does work.

Consider a cylinder of cross-sectional area $A$ closed with a weightless piston (figure 8). The work done during an expansion can be defined uniquely only if ($a$) the piston is frictionless, ($b$) the expansion is slow and always in quasi-equilibrium, i.e. the pressure outside the piston is only infinitesimally smaller than $P$ (i.e. $dP \to 0$) so that, by a minute increase in the external pressure, we would be able to *compress* the gas. Under these conditions the gas pushes with a force $PA$ against the surroundings and if the piston moves a distance $dx$ the work done is $PA\, dx = P\, dV$, where $dV$ is the volume increase. For such a *frictionless, reversible* expansion $P\, dV$ is uniquely defined for the increment $dV$.

Consider now what happens if we increase the internal energy of a gas by allowing it to do work $w$, and also add heat of amount $q$. We have

$$\overset{\text{unique}}{dU} - \overset{\text{not unique}}{-w} + \overset{\text{not unique}}{q}. \tag{2.6}$$

Here $w$ is the work done *by* the gas and is therefore counted negative. For reasons which we described before, $dU$ is determined uniquely by the initial and final states. This is not generally true of $-w$ and $q$ separately and for this reason they cannot be expressed as exact differentials.

If we now specify that the work $w$ is a reversible frictionless expansion we have

$$\overset{\text{unique}}{dU} = \overset{\text{unique}}{-P\, dV} + q. \tag{2.7}$$

Figure 8. Reversible expansion of a gas.

Consequently $q$ must also be uniquely defined over this increment of change so that it can be written as d$Q$.

Hence

$$dQ = dU + P\,dV. \tag{2.8}$$

Each of the quantities in equation (2.8) is now in the form of a differential, so if the relevant paths are known, they can all be integrated exactly. We see now why frictionless reversible expansions or compressions are so important in thermodynamics. In an electrochemical system the term equivalent to $P\,dV$ would be $E\,dq$, where $E$ is the potential difference and $dq$ the amount of charge passing. Here again for an equation like (2.8) to be valid the current flow would have to take place against an opposing potential infinitesimally less than $E$. In a magnetization process, where a magnetic element of moment $M$ is exposed to a magnetic field d$H$, the equivalent term would be $M\,dH$.

### 2.3.4 *Enthalpy*

We see from equation (2.8) that the internal energy may be written

$$dU = dQ - P\,dV. \tag{2.8a}$$

Unfortunately we cannot integrate this unless we know the equation of state and the path followed by the expansion cycle. For example an adiabatic expansion would give a value of $\int P\,dV$ which differs from that occurring in an isothermal expansion. If, however, we specify that the expansion is maintained at constant pressure we have

$$U_2 - U_1 = Q_2 - Q_1 - P(V_2 - V_1). \tag{2.9}$$

The heat absorbed at constant pressure is then given by

$$Q_2 - Q_1 = (U_2 + PV_2) - (U_1 + PV_1). \tag{2.10}$$

We now introduce a new thermodynamic quantity called enthalpy which is defined as

$$H = U + PV. \tag{2.11}$$

This is also a function solely of initial and final states. Thus for a small change

$$dH = dU + P\,dV + V\,dP = dQ + V\,dP. \tag{2.12}$$

Hence

$$\left(\frac{\partial H}{\partial T}\right)_P = \left(\frac{\partial Q}{\partial T}\right)_P. \tag{2.13}$$

As we shall see later, enthalpy is conserved in the steady flow of a gas through a throttle (Joule–Kelvin expansion).

### 2.3.5 *Specific heat capacities*

These may be defined as the amount of heat required to raise the temperature of a specified amount of material (1 mole, 1 g, 1 kg) by a unit degree of temperature:

$$C = \frac{dQ}{dT}. \tag{2.14}$$

In general we can specify two different conditions for the measurement of $C$.

Constant volume

$$C_V = \left(\frac{\partial Q}{\partial T}\right)_V = \left(\frac{dU}{dT}\right)_V \text{ from equation (2.8a).}$$

Constant pressure

$$C_P = \left(\frac{\partial Q}{\partial T}\right)_P = \left(\frac{\partial H}{\partial T}\right)_P. \tag{2.15}$$

For gases, as we shall see, there is an appreciable difference between $C_P$ and $C_V$. For liquids and solids the difference is very much smaller. Some typical values of $C$ for metals are given in Table 9. The last column shows that the product of the specific heat capacity and the atomic weight (this is in effect the specific heat capacity per mole-atom) has a value of about 25 J K$^{-1}$. This conclusion is known as the law of Dulong and Petit and we shall discuss its basis and validity in a later chapter.

Table 9. *Specific heat capacities of some metals*

| Metal | Specific heat capacity $C$ (J kg$^{-1}$ K$^{-1}$) | Atomic weight (kg) | Product (J K$^{-1}$) |
|---|---|---|---|
| Al | 920 | 0.027 | 25 |
| Cu | 380 | 0.063 | 24 |
| As | 230 | 0.108 | 25 |
| Pb | 120 | 0.207 | 25 |

### 2.3.6 *The second law of thermodynamics*

It is not the purpose of this book to provide a complete account of thermodynamics (readers will find a fascinating and stimulating account in A. B. Pippard's *Elements of Classical Thermodynamics†*). Indeed, we shall be able to deal with practically all our needs using only the first law. There is, however, some point in emphasizing the essential difference between heat and other forms of energy. When a force does work, the work done is the product of force and distance moved, the force always being measured in the direction of the line along which it is acting. When an electric current does work the current flows in the same direction as the resultant e.m.f. Mechanical work and electrical work are in principle absolutely convertible into one another. A dynamo in principle produces an amount of electrical power exactly equal to the work done in driving it (apart from friction and electromagnetic losses). If this is converted into heat there is, again, an exact equivalence as Joule showed. If, however, we wish to convert heat into work this is *not* true, because thermal energy always has associated with it random movement, vibration, thermal agitation, so there is always part of the thermal energy which is not free for conversion into mechanical or electrical energy.

N. L. S. Carnot (1824) studied the efficiency of a heat engine working between a source at a higher temperature and a sink at a lower temperature, and defined the efficiency as the ratio of useful work done to the thermal energy absorbed from the higher temperature source. He then found that the engine would achieve its greatest efficiency if all the parts of the cycle were carried out reversibly. Under these conditions the efficiency was independent of the working material. For reversible cycles, if the heat taken in is $Q_1$ and the heat rejected to the sink is $Q_2$, the efficiency $\eta$ is

$$\eta = \frac{Q_1 - Q_2}{Q_1}. \tag{2.16}$$

Since this is independent of the working material it depends only on the temperature of the two sources. This provides a means of establishing a thermodynamic scale of temperature, which is achieved by defining the ratio of heat absorbed at the higher temperature to that rejected at the lower temperature as equal to the ratio of the higher to the lower temperature, i.e. $Q_1/Q_2 = T_1/T_2$. For such an absolute scale of temperature

$$\frac{Q_1 - Q_2}{Q_1} = \frac{T_1 - T_2}{T_1}. \tag{2.17}$$

† A. B. Pippard, *Elements of Classical Thermodynamics* (Cambridge University Press, 1957).

We see that fundamentally the efficiency can never be unity unless $T_2$, the temperature of the sink, is zero. Under these conditions the heat, as it were, is acting completely in the direction of doing useful work.

For the reversible cycles we see that

$$\frac{Q_1}{T_1} = \frac{Q_2}{T_2}. \tag{2.18}$$

Hence travelling from the initial condition, through the cycle, and then back to the initial condition again

$$\frac{Q_1}{T_1} - \frac{Q_2}{T_2} = 0. \tag{2.19}$$

We may now define $dQ/T$ as the entropy change $dS$ in a reversible process. The word entropy was invented by R. Clausius; it comes from the Greek word *trope* meaning transformation and he deliberately added the prefix *en* to convey a connection with energy. For a reversible cycle (the sign $\oint$ indicates that the integral is taken over the complete cycle)

$$\oint \frac{dQ}{T} = \oint dS = 0. \tag{2.20}$$

This implies that $dS$ is an exact differential; for any increment in a reversible process

$$dS = \frac{dQ}{T} \quad \text{or} \quad dQ = T \, dS. \tag{2.21}$$

One of the most interesting properties of entropy is the following. Even for a reversible cycle the integral of $dQ$ depends on the details of the paths followed. We saw this in our discussion of enthalpy in section 2.3.4. If, however, we divide the increment of $dQ$ by the temperature at which the heat $dQ$ is absorbed (or rejected), the quantity $dQ/T$ in any reversible path can be integrated between the initial and final state. It is then found to be independent of the path followed.

Because the change in entropy is a function only of the initial and final states we can always determine the entropy change in a system even if we have brought about the change in state by irreversible or non-equilibrium processes. We merely connect the initial and final states ($A$ and $B$ respectively) by some series of reversible changes involving the addition or removal of heat and calculate the integral $\int_A^B dQ/T$. This is then the relevant entropy change. We could imagine an extreme case in which we added no heat to state $A$ but brought it to state $B$ by some type of irreversible dissipative process such as friction. In that case of course the integral $\int dQ/T$ would be zero – although the entropy change would still be that given by the reversible heating from state $A$ to state $B$. Clearly in any irreversible change the entropy change is always greater than $\int dQ/T$ if $dQ$ refers to the heat absorbed in the irreversible processes.

There are two other thermodynamic functions which are independent of path, the Gibbs free energy $G$ for isothermal changes at constant pressure and the Helmholtz free energy $A$ for isothermal changes at constant volume. They are defined as follows

$$G = U + PV - TS = H - TS, \tag{2.22a}$$

$$A = U - TS. \tag{2.22b}$$

The following simple argument (based on A. H. Cottrell's *Theoretical Structural Metallurgy†*) gives some idea of how these functions arise.

We are almost always involved in systems which exchange energy with their surroundings. However, in most cases we want to derive properties belonging to the system itself, without reference to the surroundings, in order to define the equilibrium state of the system. Suppose the system can be specified in terms of its internal energy $U$, temperature $T$, volume $V$, pressure $P$ and entropy $S$ and that it can exchange energy with its surroundings. For an infinitesimal change, using the first law we have

$$dU = dQ - P\,dV, \tag{2.8a}$$

where $dQ$ is the heat taken by the system from its surroundings and $P\,dV$ is the work it does on its surroundings. Let $dS$ and $dS_s$ be the entropy changes in the system and surroundings respectively.

The system and its surroundings constitute a closed system so that the total entropy change either increases or, in the limiting case of ideal reversibility, is zero.

$$dS + dS_s \ge 0. \tag{2.23}$$

Suppose the surroundings are at the same temperature $T$ as the system. Then since the system has taken heat $dQ$ from the surroundings the entropy change is

$$dS_s = -dQ/T. \tag{2.24}$$

Thus from equation (2.23) we have

$$dS - dQ/T \ge 0. \tag{2.25}$$

Substituting in equation (2.8a)

$$dS - (dU + P\,dV)/T \ge 0$$

or

$$dU + P\,dV - T\,dS \le 0. \tag{2.26}$$

† A. H. Cottrell, *Theoretical Structure Metallurgy* (Edward Arnold, 1948).

We now have a relation involving quantities belonging to the system itself. In equilibrium, when all changes are reversible,

$$dU + P\,dV - T\,dS = 0. \tag{2.27}$$

At constant pressure and temperature equation (2.27) may be written

$$dG = d(U + PV - TS) = 0 \text{ for constant } P, T. \tag{2.28}$$

Thus the Gibbs free energy $G$ is a minimum in the equilibrium state for changes at constant $P$ and $T$.

At constant volume and temperature $P\,dV = 0$; substituting in equation (2.26) the equilibrium condition reduces to

$$dU - T\,dS = 0$$

or

$$dA = d(U - TS) = 0 \text{ for constant } T. \tag{2.29}$$

Thus the Helmholtz free energy is a minimum in the equilibrium state for changes at constant $V$ and $T$.

There is another way of looking at these relations. Consider a constant temperature, constant volume change

$$dA = dU - T\,dS. \tag{2.30}$$

The change $dA$ is less than the change $dU$ by the quantity $T\,dS$. This is the 'unavailable' energy arising from entropy changes in the system. Thus the energy that one can extract from the system is always less than the energy available from the internal energy (except at absolute zero when $T = 0$). That is why $A$ is called the 'free' energy: part of $U$ is always entangled with the entropy factor. Similarly, at constant temperature and constant volume the maximum work that the system can perform, $dG$, is the change in enthalpy $(U + PV)$ less the unavailable energy $T\,dS$. For a non-reversible process the available energies in both cases will be less.

The unavailability of thermal energy is associated with the random nature of heat. It is not surprising therefore that entropy should in some way be connected with the probability or distribution-randomness of the system under consideration. What is surprising is that this concept can be made quantitative. If the probability of finding a system in a given state is $W_1$ and in another state is $W_2$ the entropy change may be written

$$S = k \ln \frac{W_2}{W_1}, \tag{2.31}$$

where $k$ is the Boltzmann constant. Of course, the probability needs to be correctly defined and this forms the subject of statistical mechanics.

One of the most striking examples of 'entropy in action' is the occurrence of vacancies in a crystalline solid. A finite amount of energy is required to form a vacancy so that a solid with vacancies has a higher internal energy than one without. One would therefore expect to find that, in equilibrium, a solid would tend to become vacancy-free. This is not so. Indeed the vacancy concentration increases as the temperature increases. This is because the vacancies increase the disorder: the probability distribution of atoms and vacancies is increased and this increases the entropy of the crystal. Thus, although under equilibrium conditions we are tied to zero change in the free energy

$$dA = 0 = dU - T \, dS,$$

the increase in $U$ due to vacancies is compensated for by the associated increase in $S$ (see Chapter 6). Another example of the importance of entropy is in the elasticity of rubber-like materials. This is also discussed in Chapter 6.

With gases the probability associated with the translational motion of gas molecules turns out to be proportional to the volume occupied by the gas; the larger the volume the greater the randomness. The entropy change in increasing the volume from $V_1$ to $V_2$ at constant temperature is

$$\Delta S = k \ln \left( \frac{V_2}{V_1} \right)^N,$$

where $N$ is the number of molecules in the quantity of gas considered. We shall use this result in a later chapter of the book.

# 3

## Perfect gases – bulk properties and simple theory

### 3.1 Bulk properties

#### 3.1.1 Summary of main bulk properties

As we shall see in the course of this and the next two chapters, all gases will approximate to perfect or ideal gases if they are sufficiently dilute and if the intermolecular forces are negligible compared with the thermal energy. This condition applies reasonably well to oxygen, nitrogen and hydrogen at normal temperatures and pressures. For this reason the earliest studies of the 'springiness' of air were concerned with the behaviour of a perfect gas.

The main bulk properties were established long ago. We summarize the main conclusions:

1.  Boyle's law (c. 1660). For a given mass of gas at a fixed temperature the product of pressure and volume is a constant.

$$PV = \text{constant.} \tag{3.1}$$

2.  Charles' law (1787) or Gay–Lussac's law (1802). For a given mass of gas, if the pressure be kept constant the volume increases linearly with the temperature.

$$V = V_0(1 + \alpha t). \tag{3.2}$$

If the temperature is measured on the centigrade scale and $V_0$ is the volume at $0\,°C$, it is found that $\alpha = \frac{1}{273}$. We may therefore write equation (3.2) as

$$V = V_0\left(1 + \frac{t}{273}\right) = \frac{V_0}{273}(273 + t)$$

$$= \frac{V_0}{273}\,T, \tag{3.3}$$

where $T$ is a scale of temperature, which has $-273\,°C$ as its zero point. This scale turns out to be the same as the thermodynamic scale.

3. Variation of pressure with temperature if the volume is kept constant.

$$P = P_0(1 + \beta t). \tag{3.4}$$

The constant $\beta$ has the same value of $\frac{1}{273}$ so that we have

$$P = \frac{P_0}{273} T. \tag{3.5}$$

4. General gas equation. Rewriting equation (3.1) in the light of equations (3.3) and (3.5) we obtain a general equation of state

$$PV = RT, \tag{3.6}$$

where $R$ depends only on the quantity of gas used.

5. Avogadro's law. For all perfect gases $R$ has the same value if the same molecular quantity of gas is considered. For one mole,

$$R \approx 8.3\,\mathrm{J\,K^{-1}\,mol^{-1}}.$$

6. Dalton's law. This states that, at a fixed temperature, the pressure of a mixture of gases is equal to the sum of the pressures which would be exerted by each gas separately, if the other constituents were not there.

### 3.1.2 *Reversible isothermal expansion*

We allow a gas to expand in a frictionless cylinder against a pressure infinitesimally less than the gas pressure. The temperature $T$ is maintained constant. The gas does work

$$dW = P\,dV. \tag{3.7}$$

For an ideal gas $PV = RT$ so that $P$ in equation (3.7) can be replaced by $RT/V$. Then

$$dW = \frac{RT}{V}\,dV.$$

If expansion occurs, from volume $V_1$ and pressure $P_1$ to volume $V_2$ and pressure $P_2$, the work done is

$$W = RT \ln \frac{V_2}{V_1} = RT \ln \frac{P_1}{P_2}. \tag{3.8}$$

This work is provided by the heat absorbed from the constant temperature source, which must be included in the system if $T$ is to be kept constant.

### 3.1.3 *Fast adiabatic expansion into a vacuum*

Consider the experimental arrangement shown in figure 9. A volume $V$ of gas is maintained in the left-hand flask at pressure $P$ and temperature $T$. The right-hand flask is evacuated. The whole system is thermally insulated so that heat cannot flow into or out of the flasks, then the tap is opened and the gas rushes into the empty flask. Some inequalities of temperature occur initially but a steady state is quickly reached. We write

$$dU_{total} = \Delta Q + \Delta W. \qquad (3.9)$$

For an adiabatic process there is no heat flow so that $\Delta Q = 0$. The gas expanding into the right-hand flask does work in compressing the gas as it enters the flask: but there is no net work done on or by any external system. The overall work done is zero, i.e. $\Delta W = 0$. Hence

$$dU_{total} = 0. \qquad (3.10)$$

Now the internal energy consists partly of thermal energy (which depends on temperature) and partly on potential energy due to intermolecular forces. For dry air Joule's results showed that the final temperature is indistinguishable from the original temperature. The thermal part is therefore unchanged, i.e. $dU_{thermal} = 0$. But since $dU_{total} = 0$ as in equation (3.10) it follows that $dU_{volume}$ also equals zero. This implies that there are negligible forces between the molecules.

Analytically we may write

$$U = f(V, T) \quad \text{or} \quad (P, T)$$

$$dU = \left(\frac{\partial U}{\partial V}\right)_T dV + \left(\frac{\partial U}{\partial T}\right)_V dT.$$

Since there is no temperature rise, $dT = 0$; also from equation (3.10), $dU = 0$, so consequently

$$\left(\frac{\partial U}{\partial V}\right)_T = 0, \quad \text{and similarly} \left(\frac{\partial U}{\partial P}\right)_T = 0. \qquad (3.11)$$

Figure 9. Fast adiabatic expansion of a gas into a vacuum. Joule found that for dry air there was no net temperature change.

Thus for a perfect gas, $U$ is a function of $T$ only and neither $P$ nor $V$ is involved.

### 3.1.4 *Specific heat capacity of one mole of a perfect gas*

Consider a small equilibrium change in the gas if we increase its temperature by $dT$. The basic equation is

$$dQ = dU + P\,dV$$

$$= \left(\frac{\partial U}{\partial V}\right)_T dV + \left(\frac{\partial U}{\partial T}\right)_V dT + P\,dV.$$

For a perfect gas the first term on the RHS is zero. Thus we have

$$dQ = \left(\frac{\partial U}{\partial T}\right)_V dT + P\,dV. \tag{3.12}$$

Consider now two different conditions.

(a) *Constant volume*: $dV = 0$. We can rewrite equation (3.12) as

$$dQ = \left(\frac{\partial U}{\partial T}\right)_V dT.$$

Hence

$$\left(\frac{\partial Q}{\partial T}\right)_V = C_V = \left(\frac{\partial U}{\partial T}\right)_V. \tag{3.13}$$

(b) *Constant pressure*: the gas expands during heating. Equation (3.12) now becomes

$$dQ = C_V\,dT + P\,dV.$$

Hence

$$\left(\frac{\partial Q}{\partial T}\right)_P = C_V + P\left(\frac{\partial V}{\partial T}\right)_P. \tag{3.14}$$

For a perfect gas $PV = RT$ so that $P\,dV + V\,dP = R\,dT$. At constant pressure $dP = 0$, consequently

$$P\left(\frac{\partial V}{\partial T}\right)_P = R.$$

So equation (3.14) becomes

$$\left(\frac{\partial Q}{\partial T}\right)_P = C_P = C_V + R,$$

or

$$C_P - C_V = R. \tag{3.15}$$

### 3.1.5 *Reversible adiabatic expansion*

We start with our basic equation

$$dQ = dU + P \, dV,$$

specifying that, for adiabatic process, $dQ = 0$. Our starting equation then becomes

$$0 = C_V \, dT + P \, dV. \tag{3.16}$$

For a perfect gas, the gas equation always applies:

$$P \, dV + V \, dP = R \, dT.$$

Equation (3.16) becomes:

$$- C_V \, dT = P \, dV = R \, dT - V \, dP$$
$$= (C_P - C_V) \, dT - V \, dP.$$

Hence

$$C_P \, dT = V \, dP. \tag{3.17}$$

From (3.16)

$$C_V \, dT = -P \, dV. \tag{3.18}$$

Taking the ratio of equations (3.17) and (3.18)

$$\frac{C_P}{C_V} = \gamma = \frac{- V \, dP}{P \, dV},$$

or

$$\gamma \frac{dV}{V} = -\frac{dP}{P}.$$

Integrating from initial to final conditions

$$\gamma \ln \frac{V_2}{V_1} = -\ln \frac{P_2}{P_1}$$

or

$$\ln \left( \frac{V_2}{V_1} \right)^{\gamma} + \ln \left( \frac{P_2}{P_1} \right) = 0$$

or

$$\ln \frac{P_2 V_2^{\gamma}}{P_1 V_1^{\gamma}} = 0. \tag{3.19}$$

Since the antilogarithm of 0 is 1 this gives

$$P_1 V_1^{\gamma} = P_2 V_2^{\gamma} = \text{constant.} \tag{3.20}$$

If we plot the pressure against the volume we see (figure 10) that the adiabatic curve is steeper than the isothermal.

### 3.1.6 *Work done in reversible adiabatic expansion*

The work done by the gas in expanding from $P_1 V_1$ to $P_2 V_2$ is

$$dW = P\, dV. \tag{3.21}$$

At any stage we can write

$$PV^{\gamma} = P_1 V_1^{\gamma}$$

so that $P$ in equation (3.21) can be replaced by $P_1 V_1^{\gamma} V^{-\gamma}$. The equation becomes

$$W = \int_{V_1}^{V_2} P_1 V_1^{\gamma} V^{-\gamma}\, dV.$$

Therefore

$$W = \frac{P_1 V_1}{\gamma - 1}\left[1 - \left(\frac{V_1}{V_2}\right)^{\gamma - 1}\right] = \frac{P_1 V_1}{\gamma - 1}\left[1 - \left(\frac{P_2}{P_1}\right)^{(\gamma - 1)/\gamma}\right]. \tag{3.22}$$

### 3.1.7 *Cooling as a result of adiabatic expansion*

The work done by the gas is done at the expense of the internal energy. Instead of calculating this *de novo* we merely combine the adiabatic expansion equation with the general gas equation:

$$P_1 V_1^{\gamma} = P_2 V_2^{\gamma},$$
$$\frac{P_1 V_1}{T_1} = \frac{P_2 V_2}{T_2}. \tag{3.23}$$

Figure 10. Pressure–volume relations for a perfect gas. The adiabatic (lower) curve is steeper than the isothermal (upper) curve.

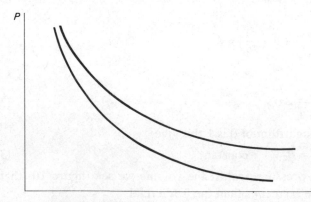

Taking the ratio of the LHS of the two equations and equating it to the ratio of the RHS we obtain

$$\frac{P_1 V_1^\gamma}{P_1 V_1/T_1} = \frac{P_2 V_2^\gamma}{P_2 V_2/T_2}.$$

Hence

$$\frac{T_2}{T_1} = \left(\frac{V_1}{V_2}\right)^{\gamma-1} = \left(\frac{P_2}{P_1}\right)^{(\gamma-1)/\gamma}. \tag{3.24}$$

### 3.1.8 *Entropy changes on expansion*

When an ideal gas expands isothermally the molecules are provided with a greater volume without any other change. This implies a greater randomness or probability and this, in turn, implies an increase in entropy.

For an isothermal expansion the entropy change for one mole of gas is easy to calculate, though the proof given here cannot be considered very rigorous. If the initial volume is $V_1$ and the final volume $V_2$ we may divide the volume into a number of unit cells, each of which could just hold one molecule. The number of cells represents the number of ways in which we could place an individual molecule in the given volume (the distribution probability). Clearly in increasing the volume from $V_1$ to $V_2$ this distribution probability for one molecule will be increased in the ratio $V_2/V_1$. If there are $N$ molecules to be considered, the distribution probability will have been increased in the ratio $(V_2/V_1)^N$. The change in entropy is then

$$\Delta S = k \ln \frac{\text{probability 2}}{\text{probability 1}} = k \ln\left(\frac{V_2}{V_1}\right)^N$$

$$= kN\left(\ln \frac{V_2}{V_1}\right) = R \ln\left(\frac{V_2}{V_1}\right), \tag{3.25}$$

if we are dealing with a mole of gas, i.e. if $N$ = Avogadro's number, $N_A$.

For the adiabatic expansion we may divide the process into two independent reversible parts. We first allow the gas to expand isothermally from $V_1$ to $V_2$, giving the entropy increase derived above. We then cool the gas (at constant volume $V_2$) until its temperature is reduced from $T_1$ to the final (adiabatic) temperature $T_2$. The decrease in entropy for such a reversible cooling is

$$\Delta S = \int \frac{dq}{T} = \int_{T_1}^{T_2} \frac{C_V \, dT}{T} = C_V \ln \frac{T_2}{T_1}. \tag{3.26}$$

But, as we saw in equation (3.24), $T_2/T_1$ for the adiabatic expansion is equal to $(V_1/V_2)^{\gamma-1}$. The entropy loss on cooling is therefore

$$C_V \ln \frac{T_2}{T_1} = C_V(\gamma - 1)\left(\ln \frac{V_1}{V_2}\right) = C_V\left(\frac{C_P}{C_V} - 1\right) \ln \frac{V_1}{V_2}$$

$$= (C_P - C_V) \ln\left(\frac{V_1}{V_2}\right)$$

$$= R \ln \frac{V_1}{V_2} = -R \ln \frac{V_2}{V_1}. \tag{3.27}$$

If we add this to the entropy change associated with the isothermal expansion part of the process (equation (3.25)), we see that the total entropy change is zero. This is because the volume increase tends to increase the randomness whereas the temperature decrease tends to reduce it, the two effects exactly balancing out. This is, of course, what we should expect, since in the adiabatic process itself no heat is added or withdrawn from the gas, so $\Delta Q = 0$ and therefore $\Delta S = \int(dQ/T)$ must be zero.

### 3.1.9 *Adiabatic lapse rate*

The pressure of the atmosphere diminishes with height and in the next chapter we shall explain why. The atmospheric pressure drops by about a tenth part for every 1 km of ascent. If at ground level the pressure is $P_0$, the pressure 1 km up is about $\frac{9}{10}P_0$.

Consider now what happens if some disturbance initiates an upward movement of air. The air will expand and in the course of this it will cool. Assuming the cooling to be purely adiabatic, ground temperature to be $T_0$, and the temperature at the 1 km level to be $T$, we have from equation (3.24)

$$\frac{T}{T_0} = \left(\frac{P}{P_0}\right)^{(\gamma-1)/\gamma} = \left(\frac{P}{P_0}\right)^{(1.4-1)/1.4} = \left(1 - \frac{1}{10}\right)^{2/7} = \left(1 - \frac{2}{70} \cdots\right),$$

$$\tag{3.28}$$

since $\gamma$ for air is approximately 1.4, If $T_0 = 300$ K,

$$T = 300\left(1 - \frac{2}{70}\right) = 300 - 9,$$

so that the temperature drop is 9 K. This is close to the observed temperature drop per km. In practice if there is an appreciable amount of moisture present in the air, cooling leads to condensation and this releases energy which lessens the cooling. A more representative value is about 6 K per km, or about 10 K per mile.

Figure 11($a$) shows the variation of temperature with height. During the day the temperature drops from about 20 °C at ground level to −60 °C at a height of 10 to 12 km, where the troposphere ends. Often in the early morning the temperature gradient is reversed near the ground level due to the cooling of the earth overnight (figure 11($b$)). Inversion temperatures of this type can be an important cause of fog and smog.

### 3.2 Elementary kinetic theory of ideal gases

The idea that gas molecules are in random motion was considered by the Greeks (as Lucretius' description of Brownian motion shows (see the end of Chapter 4)). It was revived in a surprisingly modern form by D. Bernoulli (1700–82) in 1738, but seems not to have been widely adopted until J. C. Maxwell (1831–79) took the matter in hand and provided practically the whole of our present theory of gases. We first present a very simple theory. Its basic assumptions are:

($a$) the gas consists of identical molecules of mass $m$;
($b$) they have zero size, and do not collide with one another;
($c$) they exert no forces on one another;
($d$) they undergo random motion within the gas and their collisions with the walls of the container are perfectly elastic.

Figure 11. The decrease of temperature with height in air; ($a$) steady conditions, ($b$) early morning.

We now consider the collisions of the gas molecules with the container walls and show that the pressure exerted by the gas on the walls is due to the momentum transfer accompanying the collisions.

Suppose there are $n$ molecules per m$^3$. They all have varying velocities but because the numbers are generally so enormous we can subdivide these up into groups of $n_1$, $n_2$, $n_3$ ... molecules per m$^3$ with velocity ranges $c_1$ to $c_1 + dc$; $c_2$ to $c_2 + dc$; $c_3$ to $c_3 + dc$, etc.

Let the gas be held in a cubic container of side $l$, and let us first consider the group with the velocity $c_1$ – there are $n_1 l^3$ of these in the container.

We resolve $c_1$ into three mutually orthogonal components $u_1$, $v_1$, $w_1$, parallel to the sides of the cube. Clearly

$$u_1^2 + v_1^2 + w_1^2 = c_1^2. \tag{3.29}$$

Along the $u_1$ direction the molecule has momentum $mu_1$ normal to the face of the cube before collision and momentum $- mu_1$ after collision. Thus for each collision

$$\text{momentum transfer to wall} = 2mu_1. \tag{3.30}$$

The molecule has to travel across the cube and back again (i.e. a distance of $2l$) to make the next collision. The time taken is $2l/u_1$.

$$\text{Hence the molecule makes } u_1/2l \text{ collisions per second.} \tag{3.31}$$

The momentum transfer per molecule per second is the product of equations (3.30) and (3.31), i.e.

$$2mu_1\left(\frac{u_1}{2l}\right) = \frac{mu_1^2}{l}. \tag{3.32}$$

Consequently for all the molecules of this group the total momentum transfer per second is

$$\sum \frac{m}{l} u_1^2 = \frac{m}{l} \sum u_1^2. \tag{3.33}$$

We now define the average value of $u_1^2$ for all this group as

$$\overline{u_1^2} = \frac{\sum u_1^2}{n_1 l^3}, \tag{3.34}$$

so that equation (3.33) becomes

$$\frac{m}{l} n_1 l^3 \overline{u_1^2} = mn_1 \overline{u_1^2} l^2. \tag{3.35}$$

The momentum transfer per second is the force and this is exerted on a face of area $l^2$. These molecules therefore exert a pressure $p$ on the wall of amount

$$p = mn_1 \overline{u_1^2}. \tag{3.36}$$

We now make the following observation. For this group of molecules the velocity $c_1$ is the same but their paths are uniformly distributed in all directions; when averaged over the large number of molecules involved there can be no preferred direction so that

$$\overline{u_1^2} = \overline{v_1^2} = \overline{w_1^2} = \tfrac{1}{3}c_1^2. \tag{3.37}$$

Equation (3.36) becomes

$$p = \tfrac{1}{3}mn_1c_1^2. \tag{3.38}$$

For all the groups of molecules the total pressure becomes

$$p = \Sigma \tfrac{1}{3}mn_ic_i^2$$
$$= \tfrac{2}{3}\Sigma \tfrac{1}{2}mn_ic_i^2 = \tfrac{2}{3}\text{ (total kinetic energy per m}^3\text{).} \tag{3.39}$$

For the total volume $V$ we have

$$PV = \tfrac{2}{3}\text{ (total kinetic energy of molecules in the volume }V\text{).} \tag{3.40}$$

We now define the mean square velocity of all the molecules in the gas as

$$\overline{c^2} = \frac{n_1c_1^2 + n_2c_2^2 + n_3c_3^2 + \cdots}{n_1 + n_2 + \cdots}$$
$$= \frac{\Sigma n_ic_i^2}{n}. \tag{3.41}$$

We can insert this in equation (3.39) and obtain

$$P = \tfrac{1}{3}\Sigma mn_ic_i^2 = \tfrac{1}{3}mn\overline{c^2}. \tag{3.42}$$

For a gas occupying volume $V$, where the number of molecules is $N$, where $N = nV$, we have

$$PV = \tfrac{1}{3}mN\overline{c^2}. \tag{3.43}$$

For one mole of a gas $N$ is Avogadro's number, $N_A$, and

$$PV = \tfrac{1}{3}mN_A\overline{c^2} = RT. \tag{3.44}$$

*Order of magnitude of* $(\overline{c^2})^{\frac{1}{2}}$. The square root of $\overline{c^2}$ is called the root mean square velocity. As we shall see below it is a little less than the arithmetic mean of the velocities but not very different. From equation (3.44) $RT = \tfrac{1}{3}mN_A\overline{c^2} = \tfrac{1}{3}M\overline{c^2}$, where $M$ is the molecular weight. Consequently

$$\overline{c^2} = \frac{3RT}{M}. \tag{3.45}$$

Typical values for $(\overline{c^2})^{\frac{1}{2}}$ are given in Table 10 for gases at $0\,^{\circ}$C ($T = 273$ K).

The kinetic energy of an individual molecule is $\tfrac{1}{2}mc^2$ and from (3.45) this is equal to $\tfrac{3}{2}(R/N_A)T$ or $\tfrac{3}{2}kT$, where $k$ is the Boltzmann constant. At

Table 10. *Typical values of r.m.s. velocity*

| Gas | r.m.s. velocity (m s$^{-1}$) |
| --- | --- |
| $H_2$ | 1840 |
| He | 1300 |
| $O_2$ | 650 |
| Ar | 410 |
| Benzene vapour | 290 |
| Mercury vapour | 180 |
| Electron gas | 100 000 |

room temperature the kinetic energy has a value of about $6 \times 10^{-21}$ J or 0.04 eV. It is evident that in a head-on collision between two molecules travelling in opposite directions, if all the energy could be absorbed it would be far too small to promote the internal energy of either of the molecules since the electron energies are discrete and the separations are of order 5 eV. Consequently within normal temperature ranges the kinetic energy is never converted into internal electronic energy. It is for this reason that, in the kinetic theory, we can treat the molecule as a hard sphere without any internal structure (see discussion in Chapter 1, p. 7).

### 3.3 **The ether theory of the ideal gas**

Before congratulating ourselves on the validity of the kinetic theory we may consider one of the theories which preceded it, and I am much indebted to Professor Eric Mendoza for pointing it out to me. This theory was based on the Newtonian view that a gas was like a solid. In a solid the molecules are held together by attractive forces. In a gas they are repelled by some agency which pushes them apart but still leaves them in a, more or less, uniform array. What pushes them apart? One view was that it was the swirling ether. If each molecule is surrounded by a shell of ether of radius $r$ rotating at a very high speed, the centrifugal force exerts a pressure on the neighbouring sphere and so keeps the molecules apart. It is as though the molecules are surrounded by balloons which press against one another. If the ether has a mass $m$ and the velocity is somehow in all directions, but confined to the great circles of the sphere, the net centrifugal force is $mv^2/r$ and it is exerted over an area $4\pi r^2$. The pressure exerted on the balloon shell is therefore

$$p = \frac{mv^2/r}{4\pi r^2} = \frac{mv^2}{4\pi r^3} = \frac{\frac{1}{3}mv^2}{4\pi r^3/3} .$$

The numerator on the RHS is $\frac{2}{3}$ times the kinetic energy of the ether; the denominator is the volume occupied by the ethereal shell. We are left with the relation

$$p = \tfrac{2}{3} \text{ (kinetic energy of ether per unit volume).}$$

This is very similar to equations (3.39) and shows that the right answer does not necessarily guarantee the validity of the physical model, a most salutory warning. There is, of course, more to this than coincidence. Any gas theory which attributes the gas pressure to the kinetic energy in the gas is bound to give a relation similar to that obtained in the kinetic theory. This follows because pressure has the dimensions of force/area and energy per unit volume has the dimensions of (force × distance)/volume which is also force/area.

### 3.4 Some deductions from kinetic theory
#### 3.4.1 *Thermal equilibrium between gas and container*
We now make use of a treatment described by J. Jeans in his *Kinetic Theory of Gases*† to analyse the conditions which determine the thermal equilibrium which must exist between the molecules of a gas and the molecules of the container. We start off by considering the collision between one gas molecule and one wall molecule. Let $m$, $M$ be the mass of the gas molecule and wall molecule respectively, and let their velocity components in the $x$, $y$, $z$ directions be $u$, $v$, $w$ and $U$, $V$, $W$ respectively. Suppose the collision is along the direction of $u$, $U$, so that $v$, $w$, and $V$, $W$, are unchanged. After collision, $u$ and $U$ become $u'$ and $U'$.

For an elastic collision the relative velocity is reversed so that

$$U' - u' = -(U - u). \tag{3.46}$$

The momentum is conserved, so that

$$mu + MU = mu' + MU'$$

or

$$m(u - u') = -M(U - U'). \tag{3.47}$$

Combining equations (3.46) and (3.47) we have

$$U' = \frac{1}{m + M}[(M - m)U + 2mu].$$

† J. Jeans, *Introduction to the Kinetic Theory of Gases* (Cambridge University Press, 1940).

The gain in kinetic energy of the wall molecule as a result of the collision is

$$\tfrac{1}{2}M(U'^2 - U^2) = \frac{M}{2}(U' + U)(U' - U)$$

$$= \frac{2Mm}{(m+M)^2}[mu^2 - MU^2 + (M-m)uU]. \quad (3.48)$$

If we now consider the average behaviour of a large number of collisions we note that $u$ must always be positive if collision is to occur, whereas $U$ may be positive or negative and its average value must be zero. Consequently for a large number of collisions between gas molecules and wall molecules, the average value of $uU$ is zero and the average gain of energy of each wall molecule becomes

$$\frac{2Mm}{(m+M)^2}(m\overline{u^2} - M\overline{U^2}), \quad (3.49)$$

where barred symbols refer to the mean square velocities of $u$ and $U$. For thermal equilibrium there can be no net transfer of energy to the wall. Hence

$$m\overline{u^2} = M\overline{U^2}. \quad (3.50)$$

Similar arguments may be applied to the other components $(v, V)$ and $(w, W)$. The final conclusion is that for thermal equilibrium

$$m\overline{c^2} = M\overline{C^2}, \quad (3.51)$$

where $\overline{c^2}$ and $\overline{C^2}$ are the mean square resultant velocities of the gas and wall molecules respectively. This equation implies that for thermal equilibrium the colliding molecules must have the same average kinetic energy.

This treatment assumes that the wall molecule is completely free. It is, of course, bound to the neighbouring molecules in the wall so that any analysis should really include the 'spring constants' between the wall molecules. It turns out that this does not affect the final results so that the conclusions of equations (3.50) and (3.51) remain valid.

### 3.4.2 *Boyle's law*

For a given quantity of gas

$PV = \tfrac{2}{3}$ (total kinetic energy in volume $V$).

Since the kinetic energy is constant if the temperature is constant this implies that

$PV = \text{constant}.$

### 3.4.3 *Charles' law*

If the pressure is kept constant the volume increases linearly with the kinetic energy of the molecules. Only a sophisticated thermodynamic treatment can show that the kinetic energy itself is proportional to the absolute temperature.

### 3.4.4 *Avogadro's law*

Two different gases at the same temperature must have the same mean molecular kinetic energy, since they are each individually in equilibrium with the walls of their container. But

$$PV = \tfrac{2}{3} \text{ (total kinetic energy in volume } V)$$
$$= \tfrac{2}{3} \text{ (average kinetic energy per molecule}$$
$$\times \text{ number of molecules in volume } V).$$

Hence (at constant temperature) $PV$ will have the same value for all gases provided they contain the same number of molecules.

### 3.4.5 *Dalton's law of partial pressure*

If we have two gases which separately have average molecular kinetic energies $E_1$ and $E_2$, and a concentration of $n_1$ and $n_2$ per m$^3$ respectively we may write,

$$P_1 = \tfrac{2}{3} \text{ (kinetic energy per unit volume)},$$
$$P_1 = \tfrac{2}{3} (E_1 n_1).$$

Similarly

$$P_2 = \tfrac{2}{3} (E_2 n_2).$$

For gases at equal temperatures $E_1 = E_2$, so by adding the gases together a new concentration $n = n_1 + n_2$ is obtained and the combined pressure is

$$P = \tfrac{2}{3} En = \tfrac{2}{3} E(n_1 + n_2) = \tfrac{2}{3} n_1 E_1 + \tfrac{2}{3} n_2 E_2$$
$$= P_1 + P_2. \tag{3.52}$$

### 3.4.6 *Mean free path*

We consider here the average distance between collisions of the gas molecules. Suppose $\sigma$ is the effective diameter of a molecule, then any other molecule whose centre is within a distance $\sigma$ from the molecule considered will touch it (see figure 12($a$)). For our simplest model we assume all the other gas molecules to be instantaneously at rest and follow the fate of the single mobile molecule. If this molecule has, in a given instant of time, travelled a distance $l$, it will have swept out a volume $\pi\sigma^2 l$

within which any other molecule will touch it (see figure 12($b$)). Thus if there are $n$ molecules per m$^3$ the number of molecules with which our moving molecule will collide will be $\pi\sigma^2 ln$ molecules.

The mean distance between molecular collisions $\lambda$ may now be defined as (distance travelled)/(number of collisions), i.e.

$$\lambda = \frac{l}{\pi\sigma^2 ln} = \frac{1}{\pi\sigma^2 n}. \tag{3.53}$$

This is, of course, the simplest approach possible. In practice all the molecules are not immobilized. If we allow for the true relative velocity between the molecules this reduces $\lambda$ by a factor $1/\sqrt{2}$. Also, in a collision molecules will only rarely have equi-velocity head-on collisions and bounce away from one another; usually there will be some persistence of velocity in the original direction after collision and this will complicate the analysis. Maxwell also considered the influence of intermolecular forces. All these refinements merely change the value of $\lambda$ by a small numerical factor and we shall not consider this further. Some idea of the complexity of the problem may be gathered from Jeans' book on the kinetic theory of gases.

A more important question is: how far does the concept of molecular collisions and molecular diameter affect our simple kinetic theory, where the basic assumption made was that the molecules have no size, and do not collide with one another in travelling across the container from one wall to the other? The answer is that the finite size of the molecules does influence the simple theory since the 'free' volume available for molecular movement is reduced. This will be discussed in greater detail when we consider the behaviour of 'imperfect' gases in Chapter 5. The collisions between the molecules do not, however, alter the simple theory. As we

Figure 12. ($a$) The effective diameter $\sigma$ of a molecule. ($b$) A single molecule is allowed to travel through gas molecules considered instantaneously at rest.

(a)

(b)

length of path $= l$

shall see in the next chapter, collisions deflect molecules and remove them from the group under consideration, but these are constantly replaced (as a result of other collisions) by molecules of other groups. Thus the behaviour is the same as though we could divide the molecules into non-colliding groups with specified velocities and directions.

We see from equation (3.53) that $\lambda$ varies as $1/n$, i.e. it varies as the reciprocal of the pressure $P$. Taking an average value of $\sigma = 0.3$ nm, we find that for air at s.t.p. $\lambda \approx 100$ nm, whilst the mean distance between molecules is of the order $(1/n)^{\frac{1}{3}} \approx 3$ nm. These values, 0.3, 3, 100 nm, for diameter, separation and mean free path are typical of gases under normal conditions of temperature and pressure.

### 3.4.7 *Softening of molecules*

Molecules are not hard impenetrable spheres. If the molecular speed increases they will, during collision, penetrate further into the repulsion fields of their neighbours; but this effect is very small since the repulsive forces are very powerful.

A more important effect is that which arises from the attractive or repulsive forces themselves. These will deflect the molecular paths and produce a marked change in the effective collision diameter of the molecules. An analysis for repulsive forces, falling off as the fifth power of the separation, was first given in analytical form by Maxwell. Forces of this type are not representative of most gas molecules. A more realistic model is one which considers the attractive forces which molecules exert on others passing nearby. This is illustrated schematically in figure 13. If the attractive forces are zero (figure 13($a$)), molecule B will just hit A if the separation is less than $\sigma_0$. If the attractive forces are zero and the separation is greater than $\sigma_0$, B will miss A (figure 13($b$)). If, however, attractive forces are not negligible B may well hit A even though its original direction is such that it would miss it (figures 13($b$), ($c$)). Hence, although there is no change in the size of the molecule, from the point of

Figure 13. The effect of attractive forces on the effective collision diameter.

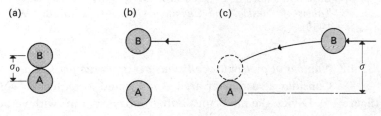

view of collision and transport phenomena, it behaves as though it had a much larger effective cross-section $\sigma$. Thus slow molecules will tend to collide even if they are very far apart, i.e. $\sigma$ tends to infinity as $T$ tends to 0. On the other hand, for very high velocities, i.e. very high temperatures, there will be practically no deflection of the molecular paths and collision will only occur when the separation is close to $\sigma_0$. A useful empirical relation for the effective molecular diameter $\sigma$ at temperature $T$ in terms of the 'geometric' diameter $\sigma_0$ is

$$\sigma^2 = \sigma_0^2\left(1 + \frac{C}{T}\right), \tag{3.54}$$

where $C$ is known as Sutherland's constant for the gas. Some typical values of Sutherland's constant are given in Table 11, and Table 12 gives values of the effective cross-section of a number of gases at temperatures from $-100\,°C$ to $300\,°C$. The effect is not large but is not negligible, especially for larger molecules.

Table 11. *Typical values of Sutherland's constant C*

| Gas | Ne | He | $O_2$ | $Cl_2$ | Steam |
|---|---|---|---|---|---|
| C (K) | 56 | 72 | 125 | 350 | 650 |

Table 12. *Effective cross-section compared with cross-section at $0\,°C$*

| Gas | $-100\,°C$ | $0\,°C$ | $100\,°C$ | $200\,°C$ | $300\,°C$ |
|---|---|---|---|---|---|
| Ne | 1.1 | 1 | 0.94 | 0.92 | 0.9 |
| $H_2$ | 1.13 | 1 | 0.94 | 0.92 | 0.9 |
| $O_2$ | 1.18 | 1 | 0.9 | 0.87 | 0.81 |
| $Cl_2$ | 1.32 | 1 | 0.85 | 0.76 | 0.70 |
| $H_2O$ | — | 1 | 0.79 | 0.68 | 0.61 |

Based on values of $C$ given in G. W. C. Kaye and T. H. Laby, *Tables of Physical and Chemical Constants*, 12th edn (Wiley, 1959).

### 3.4.8 *Number of molecular collisions per unit area per second*

Consider a surface of area $A$ subjected to molecular collisions (figure 14). Divide the molecules into groups $n_1$ per $m^3$ with velocity $c_1$;

$n_2$ per m$^3$ with velocity $c_2$, etc. Consider the first group $n_1$ and assume that they are all moving normal to the surface in question. Then in one second all the molecules in the volume $c_1A$ will strike the surface, i.e. $n_1c_1A$ molecules will strike. Hence the number of collisions per m$^2$ per second would be $n_1c_1$.

We now make use of the concept known as the Joule classification, which we used (without naming it) in our treatment of the kinetic theory. The velocity $c_1$ of a molecule may be described in terms of its $u_1, v_1, w_1$, components. Since for any one group all directions are equally possible, the number, on average, travelling in any *one* direction of given sign will be one-sixth of the total number.

Hence the collision rate m$^{-2} = \frac{1}{6}n_1c_1$.
For all the groups we have

$$\text{collisions m}^{-2}\text{ s}^{-1} = \tfrac{1}{6}\Sigma n_1c_1 = \tfrac{1}{6}n\bar{c} \tag{3.55}$$

where $n$ is the total number per m$^3$ and $\bar{c}$ is the mean velocity.

### 3.5 Transport phenomena

There are three bulk properties of gases which can be explained satisfactorily in terms of a simple molecular model, involving the transport of properties through the gas. For this reason they are called transport phenomena. Viscosity can be explained in terms of transport of momentum, heat conduction in terms of transport of thermal energy, diffusion in terms of molecular transport produced by concentration gradients.

### 3.5.1 *Viscosity*

We consider the simplest arrangement where a plane surface YY is at rest and another parallel surface ZZ at a distance $h$ away moves to the right with uniform velocity $u$. It is then found that a force $F$ must be

Figure 14. Collision of molecules travelling with velocity $c_1$ normal to a surface of area $A$.

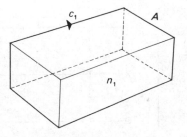

applied to the top plate to maintain the velocity $u$ against the viscous drag of the fluid between the plates (figure 15). If the flow of the fluid between the surfaces is along lines parallel to the planes YY and ZZ (lamellar flow) it is found that $F$ is proportional to the area $A$ of the moving surface and the velocity gradient $u/h$. In the more general case where the velocity gradient is not uniform across the gap, $F$ is proportional to $A$ and to $du/dy$. The factor of proportionality $\eta$ is introduced such that

$$F = \eta A \frac{du}{dy},\tag{3.56}$$

where $\eta$ is the viscosity of the liquid or gas between the surfaces.

We now show that the viscosity of a gas can be expressed in terms of momentum transported across the gas. Consider a plane XX within the gas parallel to the direction of the stream lines (figure 13($b$)). Let the velocity of flow at XX be $u$ and the velocity gradient $du/dy$. We consider the molecules entering and leaving this plane. We assume

($a$) that the flow velocity $u$ is very small compared with the mean gas velocity $\bar{c}$;

($b$) that the only molecules reaching XX are those, which, on an average, have just made their last collision at a distance $\lambda$ from XX. Then at P the molecules have a flow velocity $u + (du/dy)\lambda$ and at Q a flow velocity $u - (du/dy)\lambda$.

From assumption ($a$) the number of molecules crossing an area $A$ in one second is $\frac{1}{6}n\bar{c}A$. From Q these molecules bring to XX horizontal momentum

$$m\left(u - \frac{du}{dy}\lambda\right)\frac{n\bar{c}}{6}A.\tag{3.57}$$

Figure 15. ($a$) The concept of viscosity. ($b$) Model for describing transport of momentum in molecular terms.

From P a similar flow brings horizontal momentum

$$m\left(u+\frac{du}{dy}\lambda\right)\frac{n\bar{c}}{6}A. \tag{3.58}$$

The plane XX itself is also discharging $\frac{1}{6}n\bar{c}A$ molecules per second towards both P and Q. The resultant behaviour is summarized in Table 13.

Table 13. *Molecules entering and leaving an area A in plane XX per second*

|  | Number | Horizontal momentum |
|---|---|---|
| Entering from P | $\frac{1}{6}n\bar{c}A$ | $m\left(u+\dfrac{du}{dy}\lambda\right)\dfrac{n\bar{c}A}{6}$ |
| Entering from Q | $\frac{1}{6}n\bar{c}A$ | $m\left(u-\dfrac{du}{dy}\lambda\right)\dfrac{n\bar{c}A}{6}$ |
| Leaving from XX upwards | $\frac{1}{6}n\bar{c}A$ | $mu\dfrac{n\bar{c}A}{6}$ |
| Leaving from XX downwards | $\frac{1}{6}n\bar{c}A$ | $mu\dfrac{n\bar{c}A}{6}$ |
| Net addition to XX | 0 | 0 |

We see that there is no accumulation of molecules or momentum in plane XX. There is, however, a transport of momentum per second *through* the plane given by the difference between equations (3.58) and (3.57). This amounts to

$$\frac{n\bar{c}A}{6}\left[m\left(u+\frac{du}{dy}\lambda\right)-m\left(u-\frac{du}{dy}\lambda\right)\right]=\frac{m\lambda n\bar{c}}{3}A\frac{du}{dy}. \tag{3.59}$$

This horizontal momentum per second is a horizontal force which is transmitted through the gas to the top moving surface. As we saw before, in equation (3.56), the tangential force on the top plate is

$$F=\eta A\frac{du}{dy}.$$

By comparison we see that we have derived a relation for the viscosity of the gas

$$\eta=\tfrac{1}{3}m\lambda n\bar{c} \tag{3.60a}$$

or

$$\eta = \frac{1}{3} \frac{m}{\pi\sigma^2} \bar{c}. \tag{3.60b}$$

We conclude that the viscosity of a gas is proportional to $\bar{c}$, i.e. to $T^{\frac{1}{2}}$, and is independent of $n$, i.e. of the pressure. Both these conclusions are well supported by experiment. If, however, the pressure is too low and the mean free path becomes comparable with the distance between the moving surfaces, equations (3.60) are no longer reliable. This may be seen in the following elementary way.

Consider the transport of momentum at a distance equal to or less than $\lambda$ from the solid walls. Clearly one-half of the transport process is missing, since no molecules can pass through the walls. As there cannot be a discontinuity in the force acting across the gas we must find some way of doubling the momentum transfer. A convenient expedient is to assume that the effective velocity gradient, at a distance of about $\lambda$ from the wall, is doubled. By a very simple geometric construction shown in figure 16 we see that the slope in the bulk of the gas is then not $u/y$ but, approximately, $u/(y+2\lambda)$. Since our theory of viscosity is based on the behaviour in the bulk of the gas it is evident that the effective viscosity of the gas is reduced from the values given in equations (3.60a) and (3.60b) by a factor

$$\frac{y}{y+2\lambda}.$$

Figure 16. To ensure continuity in the force transmitted through the gas we may assume that the velocity gradient within a distance $\lambda$ from the walls, is doubled. This reduces the velocity gradient in the *bulk* of the gas from $u/y$ to, approximately, $u/(y+2\lambda)$.

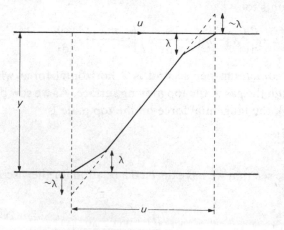

Of course a discontinuity in the velocity gradient within the gas is also inadmissible. In the more detailed analysis given below a better model is provided which explains why the factor $y/(y+2\lambda)$ is valid. The consequences are clear. If $\lambda$ is appreciable compared with $y$ the viscosity is greatly reduced. If $\lambda$ is greater than $y$ the whole analysis ceases to be valid. Finally if the molecules do not take up the tangential velocity of the wall after hitting it the transport of momentum will be less and the gas will behave as though there is slip at the walls. The more detailed analysis given below may be omitted on a first reading.

### 3.5.2 *Boundary conditions*

Suppose the distance between the plates is $y$, the lower plate is stationary and the upper plate has velocity $u$. If $y$ is very large compared with the mean free path $\lambda$, the velocity gradient across the gas (see figure $17(a)$) is

$$\frac{du}{dy} = \frac{u}{y}. \tag{3.61}$$

Figure 17. $(a)$ Ignoring the problem of the solid boundaries the velocity gradient across the bulk of the gas is $u/y$. $(b)$ If impinging molecules from the last mean-free-path separation acquire the tangential velocity of the solid wall there is an effective discontinuity in the mean tangential gas velocity close to the wall. This leads to a reduction in the velocity gradient in the bulk of the gas. $(c)$ If impinging molecules rebound 'elastically' so that their horizontal velocity component after impact is the same as before, no horizontal momentum is communicated to the walls. There is complete slipping of the molecules over the walls and zero velocity gradient in the bulk of the gas.

(a) slope $\dfrac{du}{dy} = \dfrac{u}{y}$

(b) slope $\dfrac{du}{dy} = \dfrac{u - 2\left(\frac{u_1}{2}\right)}{y}$

(c) slope $\dfrac{du}{dy} = 0$

Let us now consider in greater detail the situation at the solid plates themselves. It is clear from our derivation of $\eta$ that at the solid walls only one half of the transport process is operative since molecules cannot pass *through* the walls. Some change in flow rates must occur close to the walls if the appropriate shear stress is to be communicated through the gas to the solid wall.

Consider first the stationary wall. Molecules arriving after their last collision at a mean distance $\lambda$ away, bring with them horizontal velocity $u_1$. If all the molecules striking the wall are absorbed and then re-emitted with the tangential velocity of the (stationary) wall, this tangential velocity $u_2$ is zero (see figure 17($b$)). Thus the average velocity of the molecules very close to the stationary wall is the average of $u_1$ and zero, i.e. $u_1/2$. This is equivalent to saying that the molecules slip over the walls with velocity $u_1/2$. A similar slip occurs at the moving surface. Between these two planes at which slip occurs there is full transport of molecules to and from the plates, so that a constant velocity gradient may be maintained throughout the bulk of the gas in accordance with equation (3.61). The velocity gradient is determined by the fact that between the stationary surface and the first mean free path there is a change in velocity of $u_1/2$ over a distance $\lambda$, i.e.

$$\frac{u_1}{2} = \lambda \frac{du}{dy}.$$

The velocity difference $v$ between the gas near the stationary wall and the gas near the moving wall (see figure 17($b$)) is clearly

$$v = u - \frac{2u_1}{2},$$

or

$$v = u - 2\lambda \frac{du}{dy}.$$

Over this region there is a uniform velocity gradient of amount

$$\frac{v}{y} = \frac{(u - 2\lambda\,(du/dy))}{y} = \frac{du}{dy},$$

or

$$\frac{du}{dy} = \frac{u}{y + 2\lambda}. \tag{3.62}$$

We now see that the shear stress across every section of the bulk of the gas is communicated to the solid walls by the velocity jump close to the solid surface. Comparing equation (3.62) with (3.61), the apparent viscosity of the gas is reduced by a factor

$$\frac{y}{y+2\lambda}. \tag{3.63}$$

Clearly if the mean free path is an appreciable fraction of the separation there will be a large reduction in the apparent viscosity. In the limit, if $\lambda$ is greater than $y$, none of the concepts used in this analysis is applicable.

There is yet a further factor involved. The molecules striking the surface may not all take up the tangential velocity of the surfaces before rebounding. In the extreme case, if the tangential velocities of the molecules leaving the surface are the same as those striking it, there will be no transfer of tangential momentum to the walls. Consequently there will be no resultant velocity gradient in the bulk of the gas, but a velocity discontinuity of $u/2$ at a short distance from each of the walls (figure 17(c)). This corresponds to perfect 'slipping' of the gas over the wall surfaces without the transfer of any shear stress. Under these conditions the gas would appear to have zero viscosity. In practice the situation is much closer to that involving adsorption and re-emission with the wall velocity. As a result the simple correction, given by equation (3.63), is valid.

Finally we may refer the reader to a more quantitative treatment of this problem given by S. Chapman and T. G. Cowling in *The Mathematical Theory of Non-Uniform Gases.*† If $\theta$ is the fraction of molecules absorbed and emitted with the tangential velocity of the wall, then the fraction reflected with the same tangential velocity as the incident molecules is $1-\theta$. The average tangential velocity of the 'reflected' molecules, for example from the stationary wall, is

$$u_2 = (1-\theta)u. \tag{3.64}$$

The mean velocity at the wall is $\frac{1}{2}(u_1 + u_2)$. This is the velocity of slip. Hence

$$u_1 - \tfrac{1}{2}(u_1 + u_2) = \lambda \frac{du}{dy}. \tag{3.65}$$

Using equation (3.64) this gives for the slip velocity

$$\tfrac{1}{2}(u_1 + u_2) = \frac{2-\theta}{\theta} \lambda \frac{du}{dy}. \tag{3.66}$$

---

† S. Chapman and T. G. Cowling, *The Mathematical Theory of Non-Uniform Gases* (Cambridge University Press, 1939), Chapter 5. I am indebted to Dr N. J. Holloway and Dr B. Scruton for bringing this to my attention.

Because of similar slip at the moving wall the velocity difference across the bulk of the gas is

$$u - 2\left(\frac{2-\theta}{\theta}\right)\lambda \frac{\mathrm{d}u}{\mathrm{d}y}. \tag{3.67}$$

The apparent viscosity of the gas is thus reduced by the factor

$$\frac{y}{y + 2\lambda(2-\theta)/\theta} \tag{3.68}$$

For $\theta = 1$ this reduces to equation (3.63); for $\theta = 0$, the ratio is infinite. This corresponds to complete slip at the walls as shown in figure 17($c$).

### 3.5.3 *Thermal conductivity*

Consider heat flow across the material lying between parallel surfaces YY and ZZ (figure 18($a$)). If the temperature of the hotter surface is $T_1$ and the colder $T_2$, the rate of flow of heat is proportional to $(T_1 - T_2)/h$ and to the area of the surfaces. More generally we write for the heat flow per second

$$Q = -KA \frac{\mathrm{d}T}{\mathrm{d}y}, \tag{3.69}$$

where $K$ is the thermal conductivity of the material and $\mathrm{d}T/\mathrm{d}y$ the temperature gradient. The negative sign is because there is a flow of heat from the hotter to the colder region (positive $Q$ for negative $\mathrm{d}T/\mathrm{d}y$).

We again assume that in a gas heat conduction is due to the transport of thermal energy. If $c_V$ is the specific heat capacity at constant volume per molecule the rate of transport of thermal energy into plane XX from

Figure 18. ($a$) The concept of thermal conductivity. ($b$) Model for describing transport of thermal energy in molecular terms.

(a)          (b)

planes at a distance $\lambda$ from it is (figure 18($b$)):

from the upper plane:

$$\frac{n\bar{c}A}{6}c_V\left(T+\frac{dT}{dy}\lambda\right)$$ (3.70a)

from the lower plane:

$$\frac{n\bar{c}A}{6}c_V\left(T-\frac{dT}{dy}\lambda\right).$$ (3.70b)

Again we may show that there is no net accumulation of molecules or of thermal energy in plane XX. There is, however, a net transport of thermal energy per second of amount equal to the difference between equations (3.70a) and (3.70b).

Hence

$$Q=-\frac{n\bar{c}\lambda}{3}c_V\times A\frac{dT}{dy}.$$

By comparison with equation (3.69) we see that the thermal conductivity is given by

$$K=\frac{n\bar{c}\lambda}{3}c_V.$$ (3.71)

As in the preceding discussion of viscosity the temperature gradient near solid walls will depend on the degree of thermal equilibration between the temperature of the molecules striking the walls, and the walls themselves. If the molecules acquire the temperature of the walls, the temperature very close to the wall will show a small discontinuity, and there will be a reduction in the temperature gradient through the bulk of the gas. If the collisions are perfectly elastic so that the molecules strike the wall and are reflected with their original thermal velocity, i.e. they ignore the temperature of the wall, no heat can be transferred to or from the walls. There will be a large discontinuity in temperature at the walls: the total temperature drop will occur in a very short distance comparable with $\lambda$ and there will be a negligible temperature gradient across the mass of the gas. Consequently there will be no heat transfer. This is the limiting case in which the 'heat transfer coefficient' is zero.

*Ratio of thermal conductivity and viscosity.* From equations (3.71) and (3.60a) we have

$$\frac{K}{\eta}=\frac{c_V}{m}=\frac{c_V N_A}{m N_A}=\frac{C_V}{M}$$

where $C_V$ is the specific heat capacity per mole and $M$ the molecular weight. We therefore expect to find

$$\frac{K}{\eta}\frac{M}{C_V} = 1. \tag{3.72}$$

The experimental value of this ratio lies between 1.4 and 2.5 for a very wide range of gases. The discrepancy is due partly to molecular repulsions which tend to reduce $\eta$. A more important factor is that, in thermal conduction, the transport of the translational kinetic energy takes place more efficiently than the transfer of the rotational and vibrational forms of thermal energy. This leads to an increase in $K$ relative to $\eta$, and more sophisticated theories are available that agree well with the experimental ratio of $KM/\eta C_V$. Even the order of agreement with the simple theory provides strong support for the essential validity of the transport mechanisms described above.

### 3.5.4 *Self-diffusion*

If we have a concentration gradient in a gas, the molecules will move from the more concentrated to the less concentrated regions via molecular collisions; this involves a kinetic diffusion process. If over a distance $dx$ the concentration increases by $dn$ molecules per m$^3$ the concentration gradient is $dn/dx$. It is then found that the number of molecules per second crossing an area $A$ normal to the gradient, i.e. the rate of diffusion of molecules per second, can be written

$$\frac{dN}{dt} = -D\frac{dn}{dx}A,$$

where $D$ is called the coefficient of self-diffusion and the negative sign implies flow in the direction of smaller concentration.

Consider again a plane XX where the concentration is $n$ and neighbouring planes P and Q distant from it where the concentrations are $n + (dn/dx)$ and $n - (dn/dx)$ respectively. The number of molecules per second crossing from P

$$= \frac{1}{6}\left(n + \frac{dn}{dx}\lambda\right)\bar{c}A \tag{3.73a}$$

and from Q

$$= \frac{1}{6}\left(n - \frac{dn}{dx}\lambda\right)\bar{c}A. \tag{3.73b}$$

There are also molecules leaving on each side of XX of amount $\frac{1}{6}n\bar{c}A$. Again there is no net accumulation of molecules in XX, but the net transfer is

$$-\frac{1}{3}\lambda\bar{c}A\,\frac{dn}{dx}=-D\frac{dn}{dx}\,A.$$

Consequently

$$D=\frac{\lambda\bar{c}}{3}\quad\text{or}\quad D=\frac{\bar{c}}{3\pi n\sigma^{2}}\,. \tag{3.74}$$

It is interesting to see the orders of magnitude involved in $D$. For air at s.t.p. we have:

$$\sigma=0.25\ \text{nm};\quad\lambda=100\ \text{nm};\quad\bar{c}=450\ \text{m s}^{-1};\quad n\approx3\times10^{25}\ \text{m}^{-3}.$$

This gives a value of $D$ of order $10^{-5}\ \text{m}^{2}\,\text{s}^{-1}$. Evidently $D$ increases with the molecular velocity $\bar{c}$ but is reduced by molecular collisions as determined by $n$ and the molecular cross-section $\sigma^{2}$.

There is only one direct way of studying $D$. This is to incorporate some radioactive molecules in one specimen of gas and place it in contact with an identical gas containing no radioactive molecules. The diffusion of the radioactive species may then be followed, say, using a Geiger counter. Even here there is some difficulty since the radioactive species may not be the same size as the non-radioactive. For this reason most diffusion experiments involve the diffusion of one gas through another. If the concentration of gas 1 is $n_1$ per $\text{m}^3$ and that of gas 2 is $n_2$ per $\text{m}^3$; if $c_1$ and $c_2$ are the respective molecular velocities and $\sigma_1$ and $\sigma_2$ the respective

Figure 19. Diffusion of molecules from the surface of a liquid where vapour concentration corresponds to that of saturated vapour. The vapour condenses on a cold surface at the top of the vessel so that at this region the vapour concentration is virtually zero. This model ignores convective mass transfer.

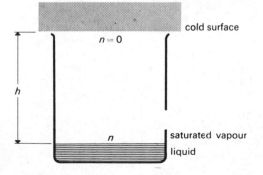

molecular cross-sections, the diffusion coefficient of gas 1 into gas 2 may be shown to be given by

$$D_{1,2} = \frac{(c_1^2 + c_2^2)^{\frac{1}{2}}}{3\pi(n_1 + n_2)(\sigma_1 + \sigma_2)^2}. \tag{3.75}$$

A typical example of $D_{1,2}$ is the diffusion of a vapour through air. Consider a cylindrical vessel of cross-sectional area $A$ (figure 19). It contains a small quantity of liquid at the bottom, and at the top open end is a cold surface on which the vapour immediately condenses. The vessel is full of air and diffusing vapour. Just above the liquid the vapour is at its saturation pressure implying a concentration $n$ molecules per m³. At the cold surface the vapour pressure is almost zero. In the steady state there can be no accumulation of vapour molecules in any section. This implies that the diffusion rate across every section is a constant. Consequently there must be a linear concentration drop from $n$ at the surface of the liquid to zero at the cold surface, distance $h$. Then

$$\frac{dn}{dx} = \frac{n}{h}.$$

Hence the mass of vapour transferred per second and condensed on the cold surface is

$$w = D_{1,2}A\frac{dn}{dx}m = D_{1,2}A\frac{n}{h}m,$$

Table 14. *Molecular diameters* $\sigma$
$(10^{-10} \text{ m})$

| Gas | Deduced from | | |
|-----|-----|-----|-----|
|     | $\eta$ | $K$ | $D$ |
| Ar | 3.6 | 3.2 | 3.8 |
| $CO_2$ | 4.5 | 3.8 | 4.5 |
| CO | 3.7 | 3.7 | 3.8 |
| $Cl_2$ | 5.4 | 5.9 | — |
| He | 2.1 | 1.9 | — |
| $H_2$ | 2.7 | 2.7 | 2.7 |
| $CH_4$ | 4.1 | 4.3 | 4.0 |
| $N_2$ | 3.7 | 3.6 | 3.7 |
| $O_2$ | 3.6 | 3.5 | 3.6 |

Adapted from G. W. C. Kaye and T. H. Laby, *Tables of Physical and Chemical Constants*, 12th edn (Wiley, 1959).

where $m$ is the weight of the molecule. The product $nm$ is the density $\rho$ of the saturated vapour. If we, therefore, determine the weight $w$ of liquid condensed on the cold surface (or lost from the liquid) per second, we have

$$D_{1,2} = \frac{wh}{A\rho}. \tag{3.76}$$

In this way it is possible to deduce values of $(\sigma_1 + \sigma_2)^2$. Some typical values of $\sigma$ deduced from measurements of $\eta$, $k$ and $D$ are given in Table 14.

### 3.5.5 *Effects of high pressures on viscosity and thermal conductivity of a gas*

We have already seen that the viscosity and thermal conductivity of a gas are independent of the pressure. This is well supported by experiment over a wide range of pressures. If, however, the pressure is so small that the mean free path becomes comparable with the separation between the surfaces the behaviour changes as indicated in a preceding section. There is also a deviation from ideal behaviour at high pressures. For pressures above a few tens of atmospheres there is a steady increase in viscosity: for example with nitrogen at a pressure of 700 atmospheres the viscosity is doubled.

The explanation depends on the major mechanism involved. If intermolecular forces are important the closer packing of the molecules will clearly increase the energy of interaction. On the other hand, if intermolecular forces are of secondary importance the closer packing itself may play an important part in increasing the viscosity. This may be seen in simple terms in the following way.

In the simple estimation of the mean free path we did not specify whether $\lambda$ was measured between centres or edges of the molecules because in general $\lambda \gg \sigma$. We can see that measuring $\lambda$ between centres is misleading when the molecules are close together: for then molecular collisions transmit impulses over a distance $\lambda + \sigma$. If we assume that this is the main factor in influencing the viscosity we may write

$$\eta = \tfrac{1}{3}nm\bar{c}(\lambda + \sigma) = \tfrac{1}{3}nm\bar{c}\lambda + \tfrac{1}{3}nm\bar{c}\sigma. \tag{3.77}$$

We note that the first term is the normal viscosity $\eta_0$: for the second term we make use of the relation $P = \tfrac{1}{3}nm\overline{c^2}$. Hence

$$\eta = \eta_0 + P\left(\frac{\bar{c}}{\overline{c^2}}\sigma\right). \tag{3.78a}$$

Ignoring the difference between $|\bar{c}|^2$ and $\overline{c^2}$ we may write

$$\eta = \eta_0 + P\left(\frac{\sigma}{\bar{c}}\right). \tag{3.78b}$$

Consequently a plot of $\eta$ against pressure $P$ should give a straight line of slope $\sigma/\bar{c}$. Data given by Jeans[†] for nitrogen have been plotted in figure 20. It is seen that except for a deviation between 1 and 200 atmospheres the points lie very close to a straight line of slope $3 \times 10^{-5} \text{ N m}^{-2} \text{ s}/ 10^8 \text{ N m}^{-2} = 3 \times 10^{-13} \text{ s}^{-1}$. This slope should equal $\sigma/\bar{c}$. For nitrogen at room temperature $\bar{c} \approx 400 \text{ m s}^{-1}$. Hence

$$\sigma = \bar{c} \times 3 \times 10^{-13} = 1.2 \times 10^{-10} \text{ m}$$
$$= 0.12 \text{ nm}.$$

Considering the crudity of the assumptions this is a reasonable value, to be compared with the more direct value of about 0.36 nm.

### 3.6 Sound waves in a gas
#### 3.6.1 *Velocity of sound in a gas*
All elastic media can transmit waves of well-defined velocity and frequency, and with solids the waves can involve shear as well as compression. But with gases the only waves that can be transmitted are

Figure 20. The effect of high pressures on the viscosity of gaseous nitrogen. Above about 50 atmospheres the viscosity increases almost linearly with pressure.

† *Introduction to the Kinetic Theory of Gases*, p. 166.

those involving a succession of compressions and rarefactions. The velocity of such a longitudinal wave is given by

$$v = \left(\frac{\text{elastic modulus}}{\text{density}}\right)^{\frac{1}{2}}.$$                    (3.79)

The elastic modulus is the bulk modulus defined as the pressure increment $dP$ divided by the fractional volume increment $dV/V$. Since a pressure increase produces a diminution in volume it is usual to define the modulus as

$$-\frac{dP}{(dV/V)}.$$

We shall soon show that sound waves are adiabatic and not isothermal. However, at this point we shall assume isothermal waves. If we write

$$PV = RT$$

then for an isothermal change $P\,dV + V\,dP = R\,dT = 0$, so that the modulus

$$-\frac{dP}{dV/V} = P.$$                    (3.80)

Hence from equation (3.79)

$$v = \left(\frac{P}{\rho}\right)^{\frac{1}{2}} = \left(\frac{nm\overline{c^2}}{3nm}\right)^{\frac{1}{2}} = \left(\frac{\overline{c^2}}{3}\right)^{\frac{1}{2}},$$                    (3.81)

since the density $\rho = nm$.
Therefore

$$v = (\overline{u^2})^{\frac{1}{2}},$$                    (3.82)

where $\overline{u^2}$ is the mean square molecular velocity in one direction. The value of $(\overline{u^2})^{\frac{1}{2}}$ is very nearly equal to $u$; this implies that the velocity of a sound wave is approximately equal to the mean molecular velocity in the direction of propagation. This at once tells us that the molecule is the messenger which carries the impulse through the gas. In this way we have linked the equation which describes wave propagation in terms of bulk properties (equation (3.79)), with a molecular process.

For an adiabatic wave the gas equation is

$$PV^{\gamma} = \text{constant}.$$

This leads to a value of the modulus

$$-\frac{dP}{dV/V} = \gamma P$$

and

$$v = \left(\frac{\gamma P}{\rho}\right)^{\frac{1}{2}}. \tag{3.83}$$

Consequently the adiabatic wave velocity is $\gamma^{\frac{1}{2}}$ times that given in equation (3.81). This increases $v$ by about 20 per cent for sound in air, but does not change the basic conclusion that sound waves are propagated by molecules communicating an impulse to neighbouring molecules by collision with them.

This conclusion is derived from detailed consideration of the equations of wave propagation. It is interesting to note that a similar view was held long before these equations were derived:

'All sounds, whether articulate or inarticulate, are produced at the meeting of bodies with other bodies or of the air with bodies, not because the air assumes certain shapes, as some people think, but because it is set in motion . . . when it clashes together by an impact from the breath or from the strings of musical instruments. For when the nearest portion of it is struck by the breath which comes in contact with it, the air is at once driven forcibly on, thrusting forward in like manner the adjoining air, so that the sound travels unaltered in quality as far as the disturbance of air manages to reach.'

*de Audibilibus*; *c.* 350 BC†

### 3.6.2 *Why sound waves are adiabatic*

It has long been a tradition in textbooks of sound to assert that in practice sound waves are adiabatic, and not isothermal, because their frequencies are so high. There is a temperature rise at the compression and a cooling at the rarefaction, and this implies that the time taken for the wave to pass along (so reversing the position of compressions and rarefactions) is not sufficient for thermal equilibrium to be reached. Consequently the wave has not the time to become isothermal. This is true but we shall now show that this is because the *frequency is relatively low* and that high frequencies (in theory) would favour *isothermal* conditions. Basically this is because the time needed for temperature equilibration to occur turns out to be proportional to the square of the wavelength, whereas the time available for this to occur is simply proportional to the wavelength. Thus equilibration occurs only if the wavelength $l$ is less than some critical length $l_c$ as indicated schematically in figure 21($a$). The following simple analysis shows how this arises.

† From *The Works of Aristotle*, vol. 6, ed. W. D. Ross (Clarendon Press, 1913). *The Oxford Classical Dictionary* suggests that *de Audibilibus* is not the work of Aristotle but of Straton, a later contemporary.

Consider an instantaneous snapshot of a plane wave travelling from left to right with velocity $v$, frequency $v$ and wavelength $l$. There are compressions at A, C, and a rarefaction at B (see figure 21($b$)). Suppose the heating at A and the cooling at B (under perfect adiabatic conditions) give a temperature difference between A and B of $T$. The temperature gradient is of course sinusoidal but for simplicity we shall assume that it is linear. Then we may write

$$\frac{dT}{dx} = -\frac{T}{\frac{1}{2}l} = -\frac{2T}{l}. \tag{3.84}$$

As a result heat flows from A to B tending to equalize the temperature, so reducing the temperature gradient. As we are only interested in an order of magnitude calculation we shall ignore such changes and assume that the gradient remains at $-2T/l$.

If $K$ is the thermal conductivity of the gas the heat flow $\Delta Q$ per m$^2$ of section in time $\Delta t$ is

$$\frac{\Delta Q}{\Delta t} = -K\frac{dT}{dx} = K\frac{2T}{l}. \tag{3.85}$$

Consequently the time $\Delta t$ taken for a quantity of heat $\Delta Q$ to flow is

$$\Delta t = \frac{\Delta Q l}{2KT}. \tag{3.86}$$

Equalization of temperature between A and B demands that the heat $\Delta Q$ raises the temperature of the mass of gas between A and B by $T$, i.e. we

Figure 21. ($a$) Time factors involved in achieving isothermal conditions in a sound wave; these are attainable only if the wavelength is less than $l_c$ ($b$) The compressions are rarefactions in a sound wave.

need an amount of heat,

$$\Delta Q = \text{volume} \times \text{number of molecules per m}^3$$
$$\times \text{specific heat capacity per molecule} \times T$$
$$= \frac{l}{2} n c_P T \text{ per unit area} .$$

(3.87)

(For an instantaneous snapshot of the wave, $c_P$ is the appropriate specific heat capacity.) Substituting this in equation (3.86) we obtain

$$\Delta t = \frac{l^2 n c_P}{4K} .$$

(3.88)

Now the time available for this equilibrium to occur cannot be greater than the time taken for the compression and rarefaction to change places, i.e. about $\tau/2$ where $\tau$ is the period of the wave. A more realistic estimate is $\tau/4 = l/4v$. Consequently temperature equilibrium will occur, i.e. we shall approximate to isothermal conditions if

$$\frac{\tau}{4} > \Delta t, \text{ i.e. } \frac{l}{4v} > \frac{l^2 n c_P}{4K} .$$

This implies

$$l < \frac{K}{n c_P v} \quad \text{or} \quad v > \frac{n c_P v^2}{K} .$$

(3.89)

In a previous section we showed (see equation (3.71)) that for a perfect gas

$$K = \frac{n \bar{c} \lambda}{3} (c_V),$$

where $\lambda$ is the mean free path. Hence isothermal conditions will be achieved if the frequency

$$v > \frac{3 n c_P v^2}{n \bar{c} \lambda c_V}, \quad \text{i.e. } v > \frac{3 v^2}{\bar{c} \lambda} ,$$

(3.90)

ignoring the difference between $c_P$ and $c_V$.

Since $v$ is of the order of $\bar{c} \approx 300 \text{ m s}^{-1}$ and $\lambda$ is about 100 nm ($10^{-7}$ m), this requires a frequency greater than $10^9$ Hz. This is extremely high. Further if we do the same substitution for $l$ we find that $l$ must be less than $\bar{c} \lambda / 3v$, i.e. it must be less than the mean free path in the gas. Under these conditions a 'normal' sound wave cannot be propagated. We conclude that sound waves are adiabatic because their frequency is not high enough for them to become isothermal. In theory they could become isothermal at frequencies of the order of $10^{10}$ Hz but such frequencies could not be transmitted through a gas as a normal sound wave.

# 4

## *Further theory of perfect gases*

In this chapter we shall first describe a rather more sophisticated kinetic theory of perfect gases. Part of the exercise is purely computational and, although it looks more impressive, adds little to our physical understanding. There are, however, a number of points which emerge which are interesting and useful and which shed new light on some of the assumptions made in the simpler forms of the kinetic theory. We shall also discuss the velocity distribution in a gas and the thermal energy of its molecules.

### 4.1 A better kinetic theory
#### 4.1.1 *Assumptions*

First we recapitulate our basic assumptions. Let us assume that we are dealing with a very large number of molecules uniformly distributed in density; that they have complete randomness of direction and velocity; that the collisions are perfectly elastic; that there are no intermolecular forces; and finally that the molecules have zero volume.

We now consider a way of describing their distribution in space. Thus to each molecule we attach a vector representing its velocity in magnitude and direction (figure 22($a$)). We then transfer these vectors (*not* the molecules) to a common origin (figure 22($b$)) and construct a sphere of arbitrary radius $r$, allowing the vectors to cut the sphere (if necessary by extending their length). Then the velocity vectors intersect the sphere in as many points as there are molecules.

If we postulate randomness of molecular motion all directions are equally probable, so that these points will be uniformly distributed over the surface of the sphere. Suppose we consider $\Delta N$ molecules, where $N$ is the total number present. These could be the total number of molecules in the vessel or the number per m³ or the number per m³ with a specified velocity range. Then the number of vector points corresponding to an element of area $\Delta A$ on the sphere will be

$$\Delta N = \frac{\Delta A}{4\pi r^2} N. \tag{4.1}$$

We can specify the element $\Delta A$ in terms of spherical coordinates $\theta$ and $\phi$ as shown in figure 23.

$$\Delta A = r \, d\theta \times r \sin \theta \, d\phi. \tag{4.2}$$

Thus the number of molecules travelling in a direction between $\theta$ and $\theta + d\theta$ and between $\phi$ and $\phi + d\phi$ (relative to some arbitrary axes) is

$$\Delta N_{\theta,\phi} = \frac{\Delta A}{4\pi r^2} N = \frac{N}{4\pi} \times \sin \theta \, d\theta \, d\phi. \tag{4.3}$$

We see that $r$ disappears from the result: $\Delta N$ depends only on the specified angles.

We could continue to use equation (4.3) in our subsequent calculations, but this would make the arithmetic more complicated than is necessary. In all the cases we are interested in, we only need to consider molecules travelling in a direction between $\theta$ and $\theta + d\theta$ irrespective of their $\phi$ position. That is to say it is sufficient for our purpose to make $\Delta A$ an annulus on the sphere lying between angles $\theta$ and $\theta + d\theta$ (figure 23(b)).

Then $\Delta A = 2\pi r \sin \theta \, r \, d\theta$, \tag{4.4}

so that the total number of molecules travelling in this direction is

$$\Delta N_\theta = \frac{\Delta A}{4\pi r^2} N = \tfrac{1}{2} \sin \theta \, d\theta \, N. \tag{4.5}$$

Figure 22. (*a*) Velocity vectors are attached to the individual molecules. (*b*) The velocity vectors are all transferred to a common origin. A sphere of arbitrary radius $r$ is drawn about the origin.

(a)                      (b)

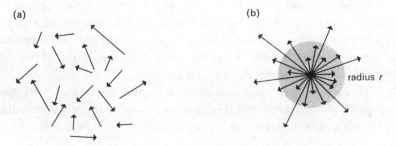

radius $r$

### 4.1.2 *Number of collisions with a solid wall*

We use the above result to calculate the number of molecules colliding per second with a square metre of a solid surface. Let us first divide the molecules into groups according to their velocities. Suppose there are $n_c$ molecules per m³ with velocity between $c$ and $c + dc$. Then the number out of this group that at any instant are travelling towards the surface from a direction $\theta$, $\theta + d\theta$ is

$$\Delta n_{c,\theta} = \tfrac{1}{2} \sin \theta \, d\theta \times n_c. \tag{4.6}$$

The only molecules that hit the surface of area $\alpha$ in time $dt$ are those contained within a prism of basal area $\alpha$ and of sloping height $c \, dt$ (see figure 24). The volume of this is $c \, dt \, \alpha \cos \theta$. Hence the number of molecules in this group hitting $\alpha$ in time $dt$ is

$$\Delta n_{c,\theta} \times c \, dt \, \alpha \cos \theta. \tag{4.7}$$

The number striking each m² per second is this number divided by $\alpha \, dt$,

$$\text{i.e. number} = \tfrac{1}{2} \sin \theta \cos \theta \, d\theta \, n_c c. \tag{4.8}$$

For all values of $\theta$ from 0 to $\pi/2$, i.e. for the whole of the 'half-space' above the surface the total collision rate is

$$\int_0^{\pi/2} \tfrac{1}{2} \sin \theta \cos \theta \, d\theta \, n_c c = \tfrac{1}{4} n_c c. \tag{4.9}$$

For all molecules of all velocity groups the number striking each m² per second is then

$$\tfrac{1}{4} \Sigma n_c c = \tfrac{1}{4} n \bar{c}, \tag{4.10}$$

Figure 23. Specification of an element of area on the velocity sphere (*a*) in terms of angles $\theta$ and $\phi$. (*b*) in terms only of angle $\theta$.

(a)                                                      (b)

where $n$ is the total number per m$^3$ and $\bar{c}$ is the mean velocity defined by $\bar{c} = \Sigma n_c c / n$. This result differs from that given in the previous chapter (equation (3.55)) where the fraction in front of $n\bar{c}$ was found to be $\frac{1}{6}$ instead of $\frac{1}{4}$.

### 4.1.3 *Gas equation of state*

We now consider the momentum transfer per second per m$^2$ of wall, i.e. the pressure exerted by the gas. We first consider the group of molecules $n_c$ per m$^3$ whose velocity is between $c$ and $c + dc$. Of these consider the group approaching the surface at a direction $\theta$, $\theta + d\theta$; each molecule brings with it momentum $mc$. The vertical component is, therefore, $mc \cos \theta$ and after collision it is reversed so that the net momentum transferred to the wall is $2mc \cos \theta$. The horizontal component $mc \sin \theta$ is unchanged so that no tangential momentum is communicated to the wall. The vertical momentum change per second transferred to each m$^2$ of wall, i.e. the pressure, is

$$2mc \cos \theta \times (\text{number hitting each m}^2 \text{ per second})$$
$$= 2mc \cos \theta \, \tfrac{1}{2} \sin \theta \cos \theta \, d\theta \, n_c c$$
$$= m n_c c^2 \cos^2 \theta \sin \theta \, d\theta. \tag{4.11}$$

Integrating over the whole half-space we obtain the pressure produced by this group of molecules:

$$m n_c c^2 \int_0^{\pi/2} \cos^2 \theta \sin \theta \, d\theta$$
$$= m n_c c^2 \left[ -\frac{\cos^3 \theta}{3} \right]_0^{\pi/2} = \tfrac{1}{3} m n_c c^2. \tag{4.12}$$

For all groups of molecules we thus obtain

$$p = \tfrac{1}{3} m n \overline{c^2}, \tag{4.13}$$

where the mean square velocity $\overline{c^2}$ is given by $\Sigma n_c c^2 / n$.

Figure 24. The number of molecules hitting the area $\alpha$ in time $dt$ is equal to the number of molecules within the prism of basal area $\alpha$ and sloping height $c \, dt$.

### 4.1.4 *Transport phenomena*

Consider for example, the viscosity of a gas in terms of the momentum transport across a plane as described in the previous chapter. The flow is assumed streamlined parallel to the XX plane, the velocity gradient $du/dy$ in the steady state is constant. We have to calculate the transport of horizontal momentum across each m$^2$ in the plane XX per second. Consider a small area element $d\alpha$ in XX. The only molecules we are interested in are those which on an average made their last collision at a distance $\lambda$ from the element. Draw a hemisphere of radius $\lambda$ with the element as centre (see figure 25). Those coming from the direction $\theta$, $\theta + d\theta$, are at a distance $\lambda \cos \theta$ from XX so that these molecules bring additional horizontal momentum $m\lambda \cos \theta \, (du/dy)$. The number coming from this direction and striking each m$^2$ of the element per second is given by equation (4.8). The transfer of horizontal momentum per m$^2$ per second is therefore:

$$m\lambda \cos \theta \left(\frac{du}{dy}\right) \times \tfrac{1}{2} \sin \theta \cos \theta \, d\theta \, n_c c. \tag{4.14}$$

If we integrate this over the whole of the top half-space we obtain

$$\tfrac{1}{6} m\lambda n_c c \times \left(\frac{du}{dy}\right). \tag{4.15}$$

There is a similar contribution from the half-space below so that the tangential stress in the plane XX is simply

$$\tfrac{1}{3} m\lambda n_c c \left(\frac{du}{dy}\right). \tag{4.16}$$

If we equate this to the viscous stress $\eta(du/dy)$ we obtain

$$\eta = \tfrac{1}{3} mn_c c\lambda, \tag{4.17}$$

Figure 25. Transport of momentum through element of area $d\alpha$. A hemisphere of radius $\lambda$ is constructed with $d\alpha$ as centre. All molecules from this hemisphere have, on average, just made their last collision before proceeding towards $d\alpha$ from half of the gas.

for molecules of this group. For all groups of molecules

$$\eta = \tfrac{1}{3}mn\bar{c}\lambda. \tag{4.18}$$

The result is identical with that obtained from the Joule classification.

We may explain this result in a simple way. The number of molecules striking each m$^2$ in XX per second is $\tfrac{1}{4}n\bar{c}$ instead of $\tfrac{1}{6}n\bar{c}$ as in the Joule derivation, but, whereas in the Joule derivation each molecule is assumed to come from a constant distance $\lambda$ from plane XX, in the present treatment the last collision occurs on the hemisphere of radius $\lambda$, so that the last collision distance measured normal to XX ranges from 0 to $\lambda$. The average effective distance is $\tfrac{2}{3}\lambda$. The horizontal momentum transfer is thus the same in both models.

### 4.1.5 *Mean free path $\lambda$ and collision frequency $v$*

We now extend our treatment to allow for the finite diameter $\sigma$ of the molecules. As we saw in the previous chapter, a very simple result for the average distance $\lambda$ between each collision is given by

$$\lambda = \frac{1}{\pi\sigma^2 n}. \tag{4.19}$$

There are many sophisticated ways of improving this relation but they only change $\lambda$ by a small numerical factor.

We now define the collision frequency $v$ as the average number of collisions per second made by each molecule.

Then

$$v = \frac{\text{average distance travelled per second}}{\text{average distance between collisions}},$$

$$v = \frac{\bar{c}}{\lambda}. \tag{4.20}$$

This result does not depend on the way in which $\lambda$ is derived.

### 4.1.6 *Distribution of free paths: survival equation*

We wish to follow the fortunes of a group of $N$ molecules. As they travel they collide with themselves and other molecules. Can we estimate the number that, at any specified stage, have not yet made a collision?

Suppose that, at some instant, $n$ have survived without collision. If each is allowed to travel a further distance d$x$ along its free path further collisions may occur. We *assume* that this number of collisions is proportional to both $n$ and d$x$. Hence the number removed by these collisions is proportional to $n\,\mathrm{d}x$. To show that $n$ is decreased we write:

$$\mathrm{d}n = -Pn\,\mathrm{d}x, \tag{4.21}$$

where $P$ is the collision probability for the gas. We assume this to be a constant for a given gas under specified conditions. Hence,

$$\frac{\mathrm{d}n}{n} = -P\,\mathrm{d}x,$$

$$n = N\,\mathrm{e}^{-Px} \tag{4.22}$$

and

$$\mathrm{d}n = -PN\,\mathrm{e}^{-Px}\,\mathrm{d}x. \tag{4.23}$$

Since $n$ is the number surviving after travelling a distance $x$, the number $n_c$ which have suffered collision (see figure 26) is simply

$$n_c = N - n.$$

Hence

$$\mathrm{d}n_c = \mathrm{d}(N-n) = -\mathrm{d}n,$$

or

$$\mathrm{d}n_c = +PN\,\mathrm{e}^{-Px}\,\mathrm{d}x.$$

Figure 26. Graph showing survival of an initial group of $N$ molecules. After travelling a distance $x$, $n$ have survived without collision whilst the number $n_c$ suffering collision is given by $n_c = N - n$.

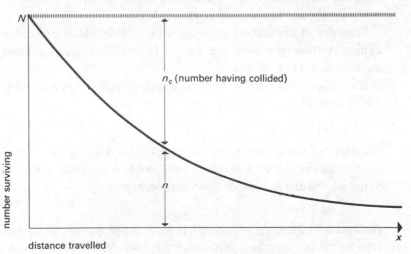

This is the number of molecules with mean free path between $x$ and $x + dx$. The average mean free path $x$ is given by

$$\lambda = \frac{\int_{n=0}^{n=N} x \, dn_c}{\int_0^N dn_c},$$

i.e.

$$\lambda = \frac{\int_{x=0}^{x=\infty} xPN \, e^{-Px} \, dx}{N}$$

$$= -\left[ x \, e^{-Px} + \frac{e^{-Px}}{P} \right]_0^\infty = \frac{1}{P}.$$

Consequently, equation (4.22) becomes

$$n = N \, e^{-x/\lambda}. \tag{4.24}$$

This is the basic 'survival equation', where $n$ is the number of molecules out of a group $N$ that have not yet made a collision after travelling a distance $x$.

### 4.1.7 *Collisions with surfaces: a final treatment*

We may now ask a simple but important question in relation to the kinetic theory as we have so far developed it. We have always assumed that if somewhere there is a group of molecules travelling in some specified direction we can calculate the number striking $1 \, m^2$ of a surface per second by completely ignoring intermolecular collisions. The previous section shows that this is quite unjustified. Molecules a long way from the surface never get there: they are diverted by collisions on the way with themselves and other molecules.

Consider at any instant an element of volume $dV$ in a gas of uniform density containing $n$ molecules per $m^3$ (figure 27($a$)). The number in the element is $n \, dV$.

If $\nu$ is the collision frequency, the number of collisions occurring within $dV$ in time $dt$ is

$$\tfrac{1}{2}\nu(n \, dV) \, dt. \tag{4.25}$$

The $\frac{1}{2}$ is introduced since otherwise each collision would be counted twice. Each collision starts off two new free paths so the total number of free paths originating in time $dt$ from the element $dV$ is

$$\nu n \, dV \, dt.$$

Because all directions of molecular motion are equally probable in any element of the gas, these molecules start off uniformly in all directions.

The fraction heading towards the area $\Delta A$ is $\Delta\omega/4\pi$, where $\Delta\omega$ is the solid angle subtended by $\Delta A$ at the volume element.

Clearly,

$$\Delta\omega = \frac{\Delta A \cos\theta}{r^2}. \qquad (4.26)$$

The number heading towards $\Delta A$ is thus

$$\Delta n = \frac{\Delta\omega}{4\pi} \times \nu n \, dV \times dt. \qquad (4.27)$$

However, from the survival equation, the number of molecules reaching $\Delta A$ without having made a collision, and so not having been eliminated is

$$\Delta n_1 = \Delta n \, e^{-r/\lambda}.$$

Consequently, the number of molecules leaving $dV$ in time $dt$ and ultimately reaching $\Delta A$ without having made an intermediate collision is

$$\Delta n_1 = \frac{1}{4\pi} \frac{\Delta A \cos\theta}{r^2} \nu n \, e^{-r/\lambda} dV \, dt. \qquad (4.28)$$

We note that $\Delta n_1$ depends on the time element $dt$, not on the absolute time. This is because on the average any given element of volume appears unchanged. (To some extent this observation, in itself implies the conclusion we shall draw in the following paragraphs.) We may thus choose our time element $dt$ for each volume element $dV$ at various times, but in such a way that all the undeflected molecules from all the volume elements arrive at $\Delta A$ during the same time interval $dt$. This enables us to integrate for the whole volume of gas above $\Delta A$. Once again we choose as

Figure 27. (a) Molecules in element of volume $dV$, after their last collision start travelling towards $\Delta A$. (b) Conical annulus of volume $2\pi r \sin\theta r \, d\theta \, dr$ defines $dV$.

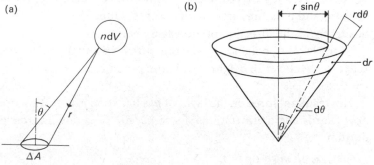

our element of volume a conical annulus of radius $r \sin \theta$, thickness $r \, d\theta$, height $dr$ (figure 27($b$)), so that

$$dV = 2\pi r \sin \theta \, r \, d\theta \, dr. \qquad (4.29)$$

We substitute in equation (4.28) and integrate for $r = 0$ to $r = \infty$, and for $\theta = 0$ to $\theta = \pi/2$. This gives the total number of molecules, from all half-space, from all directions striking $\Delta A$ in time $dt$ as:

$$\nu n \, \Delta A \, dt \int_0^\infty \int_0^{\pi/2} \frac{\cos \theta \sin \theta}{2} \, d\theta \, e^{-r/\lambda} \, dr \qquad (4.30)$$

$$= \tfrac{1}{4} \nu n \lambda \, \Delta A \, dt.$$

The number striking per m$^2$ per second is then

$$\tfrac{1}{4} \nu n \lambda. \qquad (4.31)$$

But, irrespective of our model for calculating $\lambda$,

$$\nu = \frac{\bar{c}}{\lambda}.$$

Hence the number striking each m$^2$ per second is

$$\tfrac{1}{4} n \bar{c}. \qquad (4.32)$$

This is the same as the result we have already obtained. It implies that although molecules are continuously colliding and changing their directions and velocities, other molecules are replacing the ones that are eliminated. For this reason the simple picture which ignores intermolecular collisions in calculating wall collisions, gas pressure and transport phenomena yields the correct answer.

## 4.2 Sedimentation
### 4.2.1 *Sedimentation under gravity*
Consider a gas column of unit cross-section and at uniform temperature $T$ (figure 28). We shall show that because of gravity there is a density (and pressure) gradient in the gas. Our increments are measured positively in the direction of increasing height $h$.

At the level A, height $h$, let the pressure be $p$ and the molecular density $n$ per m$^3$. At level B, height $h + dh$, the pressure is $p + dp$ and the density $n + dn$.

The resultant force on the layer of gas between A and B is a pressure $dp$ downwards and a gravitational force $n \times dh \times mg$ downwards. For equilibrium

$$n \, dh \, mg + dp = 0. \qquad (4.33)$$

Since $p = \frac{1}{3}nm\overline{c^2}$ and $\overline{c^2}$ is constant at all levels (because of constant temperature)

$$dp = \frac{1}{3}m\overline{c^2}\,dn.$$

Equation (4.33) becomes

$$\frac{dn}{n} = \frac{-mg}{\frac{1}{3}m\overline{c^2}}\,dh.$$

Integrating from $h = 0$ to $h$ and from $n_0$ to $n$ we obtain

$$\ln\frac{n}{n_0} = \frac{-mg}{\frac{1}{3}m\overline{c^2}}\,h = \frac{-Nmg}{\frac{1}{3}Nm\overline{c^2}}\,h,$$

or

$$\ln\frac{n}{n_0} = \ln\frac{p}{p_0} = -\frac{Mg}{RT}\,h. \qquad (4.34)$$

For air, substituting $M \approx 0.030$ kg, $R = 8.3$ J, $T = 300$ K, we find that at a height of 1 km the pressure is reduced by about one-tenth of an atmosphere.

### 4.2.2 *Sedimentation of particles*

If particles of mass $m$ and density $\rho$ are suspended in a liquid of density $\rho_0$, then if there is complete thermal equilibrium, the concentration of particles is given by

$$\ln\frac{n}{n_0} = -\frac{Nmgh}{RT}\frac{(\rho - \rho_0)}{\rho}. \qquad (4.35)$$

An examination of the variation of $n$ with height for a fine suspension was used by J. Perrin in 1908 to calculate $N_A$. He obtained a value of $6 \times 10^{23}$, which is virtually identical with the value of Avogadro's number obtained

Figure 28. Decrease of density and pressure with height of an ideal gas.

by other methods. This shows that fine particles in thermal equilibrium behave like a gas of molecular mass $m$. It is quite wrong to imagine that this is simply because they are buffeted around by the molecules of the liquid in which they are suspended. They possess this property simply by virtue of being in thermal equilibrium with themselves and their surroundings. Of course the collisions between the particles and the liquid molecules ensure that the whole system is in thermal equilibrium, but the kinetic theory applies to the particles themselves just as it does to the molecules of a conventional gas. Equation (4.35) also shows that by increasing $g$ the major fraction of the particles will be formed in a thin layer near the bottom of the vessel. This is the basis of the action of a centrifuge.

### 4.2.3 *The Boltzmann distribution*

We may rewrite equation (4.34) in the form

$$\ln \frac{n}{n_0} = -\frac{Mg}{RT}h = -\frac{mg}{kT}h,$$

where $k = R/N =$ the Boltzmann constant. Then

$$n = n_0 \exp\left(-\frac{mg}{kT}h\right). \tag{4.36}$$

Since the temperature is constant throughout the gas the kinetic energy is constant at all levels; only the potential energy has been changed, by amount $mgh$. Then $mgh$ is the amount by which the energy of a molecule at $h$ exceeds that at ground level; call this $\varepsilon$. Then

$$n = n_0 \exp\left(-\frac{\varepsilon}{kT}\right). \tag{4.37}$$

Boltzmann showed that for any equilibrium distribution of molecules this relation generally holds. We may write

$$\begin{pmatrix} \text{number of molecules at energy} \\ \text{level } \varepsilon \text{ above ground level} \end{pmatrix} = \begin{pmatrix} \text{number of molecules} \\ \text{at ground level} \end{pmatrix} \times \exp\left(-\frac{\varepsilon}{kT}\right).$$

Again for two energy states $\varepsilon_1$ and $\varepsilon_2$

$$\frac{n_1}{n_2} = \frac{\exp(-\varepsilon_1/kT)}{\exp(-\varepsilon_2/kT)} = \exp\left(-\frac{\Delta\varepsilon}{kT}\right), \tag{4.38}$$

where $\Delta\varepsilon$ is the energy gap between states 1 and 2. The Boltzmann function thus describes the relative population of energy states in an equilibrium system.

### 4.3  **Temperature variation of reaction rates**

A chemical reaction can only occur spontaneously if the final system has a lower free energy state than the initial system. In general, however, before the reaction can take place one or more of the reactants must be excited to a higher energy state (see figure 29). The rate of forward reaction is then determined by the number of molecules which have enough energy to get over this barrier. The forward reaction rate is proportional to $\exp(-\varepsilon_f/kT)$, where $\varepsilon_f$ is the activation energy.

It is interesting to note that the back reaction rate will be proportional to $\exp(-\varepsilon_b/kT)$, and that the heat of reaction is equal to the difference between the activation energies of forward and backward reaction,

$$q = \varepsilon_b - \varepsilon_f.$$

### 4.4  **Distribution of velocities in a perfect gas**

#### 4.4.1  *Velocity distribution in a one-, two- and three-dimensional gas*

We expect to find a wide distribution of molecular velocities in a gas, because even if we could start off with all the molecules travelling in different directions with equal velocities, random collisions are bound to speed some and retard others. Consequently we expect to find the velocities ranging from 0 to ∞ but grouped around a definite average

Figure 29. Forward reaction has an activation energy $\varepsilon_f$, the backward reaction an activation energy $\varepsilon_b$. The heat of reaction is $q = \varepsilon_b - \varepsilon_f$.

(a)                                        (b)

value determined by the temperature.† The velocity distribution is, in fact, of the same type as the Boltzmann function.

If there are no intermolecular forces the only energy the individual molecule has is its kinetic energy $\frac{1}{2}mc^2$. (We ignore here vibrational or rotational energies which are assumed to be unchanged by the gas-collision process.) Then according to the Boltzmann distribution the population density of molecules with velocity $c$ is proportional to

$$\exp\left(-\frac{mc^2}{2kT}\right). \tag{4.39}$$

If we attach velocity vectors to the molecules and transpose them to a single origin (see figure 30(a)), the number of velocity vectors ending in the element $du$, $dv$, $dw$ gives us the number of molecules with velocities between $u$ and $u + du$, $v$ and $v + dv$, $w$ and $w + dw$. This number is then

$$dn = A \exp\left(-\frac{mc^2}{2kT}\right) du \, dv \, dw, \tag{4.40}$$

where $A$ is a suitable constant.

Before proceeding to use this relation we may split up $c^2$ into its three components $u^2 + v^2 + w^2$ so that

$$dn = A \exp\left(-\frac{mu^2}{2kT}\right) \exp\left(-\frac{mv^2}{2kT}\right) \exp\left(-\frac{mw^2}{2kT}\right) du \, dv \, dw. \tag{4.41}$$

Since each of the three components must have the same velocity distributions, the quantity

$$A^{\frac{1}{3}} \exp\left(-\frac{mu^2}{2kT}\right) du$$

gives the number of molecules with velocity between $u$ and $u + du$ irrespective of the $v$ and $w$ components. The fraction of molecules in this category is

$$\frac{dn}{n} = B \exp\left(-\frac{mu^2}{2kT}\right) du. \tag{4.42}$$

The constant $B$ may be determined by observing that, if this relation is integrated for all possible velocities from $-\infty$ to $+\infty$, the resulting

† Because of relativity effects an infinite velocity would imply an infinite effective mass. However, the fraction of molecules with velocities high enough to produce appreciable relativity effects is negligible. In what follows we shall assume that relativity effects may be ignored and that the velocity distribution extends to infinity although, of course, infinite velocities cannot in fact occur.

fraction must be unity. Then

$$\int_{-\infty}^{\infty} B \exp\left(-\frac{mu^2}{2kT}\right) du = 1. \qquad (4.43)$$

Putting

$$\frac{m}{2kT} = a, \qquad (4.44)$$

$$B \int_{-\infty}^{\infty} e^{-au^2} du = 1.$$

From the table of integrals on p. 116,

$$\int_{-\infty}^{\infty} e^{-au^2} du = 2 \int_{0}^{\infty} e^{-au^2} du = \left(\frac{\pi}{a}\right)^{\frac{1}{2}} = \left(\frac{2\pi kT}{m}\right)^{\frac{1}{2}}.$$

Hence

$$B = \left(\frac{m}{2\pi kT}\right)^{\frac{1}{2}}.$$

Consequently

$$\frac{dn}{n} = \left(\frac{m}{2\pi kT}\right)^{\frac{1}{2}} \exp\left(-\frac{mu^2}{2kT}\right) du. \qquad (4.45)$$

How can we plot this function? We bring the term $du$ over to the LHS so that we have

$$\frac{dn}{n}\frac{1}{du} \quad \text{or} \quad \frac{1}{n}\frac{dn}{du} = \left(\frac{m}{2\pi kT}\right)^{\frac{1}{2}} \exp\left(-\frac{mu^2}{2kT}\right). \qquad (4.46)$$

We may calculate the RHS and plot it as a function of $u$ (see figure 30($b$)). If we take a strip of width $du$ its area is then $dn/n$ which is the fraction of

Figure 30. Velocity vector diagrams ($a$) for molecules with velocities between $c$ and $c+du+dv+dw$, ($b$) velocity distribution for a one-dimensional gas. The area of the shaded band gives the fraction of molecules with velocities between $u$ and $u+du$ in the $u$ direction.

(a)  (b)

molecules with velocity between $u$ and $u + du$. We see that the greatest fraction occurs for zero velocity; this is because of the exponential nature of the Boltzmann factor, and the fact that the 'one-dimensional' gas directly reflects this distribution.

The average energy in the $u$ direction is given by

$$\bar{\varepsilon} = \left[ \int_{-\infty}^{\infty} \left( \frac{mu^2}{2} \right) dn \right] \frac{1}{n}$$

$$\bar{\varepsilon} = \frac{m}{2} \left( \frac{m}{2\pi kT} \right)^{\frac{1}{2}} \int_{-\infty}^{\infty} u^2 e^{-au^2} du = \frac{m}{2} \left( \frac{a}{\pi} \right)^{\frac{1}{2}} \int_{-\infty}^{\infty} u^2 e^{-au^2} du.$$

(4.47)

From the table of integrals on p. 116

$$\int_{-\infty}^{\infty} u^2 e^{-au^2} du = \frac{1}{2} \left( \frac{\pi}{a^3} \right)^{\frac{1}{2}}.$$

Hence

$$\bar{\varepsilon} = \frac{m}{2} \left( \frac{a}{\pi} \right)^{\frac{1}{2}} \frac{1}{2} \left( \frac{\pi}{a^3} \right)^{\frac{1}{2}} = \frac{m}{4} \times \frac{1}{a} = \frac{m}{4} \frac{2kT}{m},$$

or

$$\bar{\varepsilon} = \frac{kT}{2}.$$

(4.48)

Similarly the fraction of molecules with velocity components between $u$ and $u + du$ and between $v$ and $v + dv$ is simply the product of the two individual probabilities.

$$\frac{dn}{n} = \left( \frac{m}{2\pi kT} \right) \exp \left[ - \frac{m(u^2 + v^2)}{2kT} \right] du \, dv.$$

(4.49)

Finally the fraction of molecules with velocity components between $u$ and $u + du$, $v$ and $v + dv$, $w$ and $w + dw$ is

$$\frac{dn}{n} = \left( \frac{m}{2\pi kT} \right)^{\frac{3}{2}} \exp \left[ - \frac{m(u^2 + v^2 + w^2)}{2kT} \right] du \, dv \, dw.$$

(4.50)

Generally we are not interested in the dependence of $n$ on the individual components of velocity. For practically all purposes we wish rather to know the number of molecules with velocities between $c$ and $c + dc$, irrespective of direction. We may calculate this easily by drawing a sphere of radius $c$ and another of radius $c + dc$ (see figure 31(a)). The number of velocity vectors ending in the spherical shell is then the number required. In velocity space the volume of the shell is $4\pi c^2 \, dc$ and this replaces $du \, dv \, dw$ in equation (4.50). We then obtain

$$\frac{1}{n} \frac{dn}{dc} = 4\pi c^2 \left( \frac{m}{2\pi kT} \right)^{\frac{3}{2}} \exp \left( - \frac{mc^2}{2kT} \right).$$

(4.51)

This is plotted in figure 31($b$). We note that in any integration to give total quantities, $c$ ranges from zero to $\infty$, not from $-\infty$ to $+\infty$.

[The reader, if he is interested only in the three-dimensional gas, may pass at once from equation (4.50) to equation (4.51) by writing equation (4.40) in the form

$$\frac{dn}{n} = C \exp\left(-\frac{mc^2}{2kT}\right) 4\pi c^2 \, dc.$$

By specifying that $\int_{c=0}^{c=\infty} (dn/n) = 1$, he will find that $C$ has the value $(m/2\pi kT)^{\frac{3}{2}}$. This results in equation (4.51).]

### 4.4.2 *Average velocities*

Rewriting equation (4.51) in the form

$$\frac{dn}{n} = 4\pi \left(\frac{a}{\pi}\right)^{\frac{3}{2}} e^{-ac^2} c^2 \, dc, \tag{4.52}$$

and using, if required, the integrals on p. 116, we may calculate the following three representative velocities: the most probable velocity $c_m$, i.e. the velocity for which equation (4.52) is a maximum, the average velocity $\bar{c}$, and the root mean square velocity $(\overline{c^2})^{\frac{1}{2}}$. We have:

most probable

$$c_m = \sqrt{2}\left(\frac{kT}{m}\right)^{\frac{1}{2}}, \tag{4.53}$$

Figure 31. ($a$) Velocity vector diagram for molecules with velocities between $c$ and $c + dc$. This is given by the numbers of points ending the velocity vectors which lie within the spherical shell of radii $c$ and $c + dc$. ($b$) Maxwellian velocity distribution for a three-dimensional gas. The shaded band gives the fraction of molecules with velocities between $c$ and $c + dc$. The approximate positions of the most probable velocity $c_m$ the mean velocity $\bar{c}$ and the root mean square velocity $(\overline{c^2})^{\frac{1}{2}}$ are indicated.

(a)  (b)

average

$$\bar{c} = \int_0^\infty c\,\frac{dn}{n} = \left(\frac{8}{\pi}\right)^{\frac{1}{2}}\left(\frac{kT}{m}\right)^{\frac{1}{2}},\tag{4.54}$$

r.m.s.

$$(\overline{c^2})^{\frac{1}{2}} = \left(\int_0^\infty \frac{c^2\,dn}{n}\right)^{\frac{1}{2}} = \sqrt{3}\left(\frac{kT}{m}\right)^{\frac{1}{2}}.\tag{4.55}$$

Before leaving this we note one of the confusing features of probability distributions. According to equation (4.39) the population density is a maximum when $c = 0$. This is also true, as we saw above of the velocity distribution in a 'one-dimensional' gas. As soon, however, as we specify velocity elements in two or three dimensions the most probable velocity (as distinct from the probability density) no longer corresponds to zero velocity. This is specifically brought out in equation (4.51) and figure 31 for a three-dimensional gas. A similar point will arise when we discuss the behaviour of rubber molecules in Chapter 7.

### 4.4.3 *Number of molecules striking a surface per second*

The number of molecules striking a square metre of surface per second may be directly obtained from equation (4.45). If the surface considered is at right angles to the $u$ direction the $v$ and $w$ components are not involved. Consider molecules with velocity $u$, $u + du$; we construct a prism of unit cross-section and length $u$. Then all the molecules in this volume will reach the surface in one second. This number is $u\,dn$ where $dn$ is the number per m$^3$ with velocity $u$ in the sense *towards* the surface. The total number required is the integral of $u\,dn$ for all values of $u$ from 0 to infinity: this choice of limits ensures that we are considering only those molecules moving towards the surface. The number is

$$n\int_0^\infty \left(\frac{m}{2\pi kT}\right)^{\frac{1}{2}} \exp\left(-\frac{mu^2}{2kT}\right) u\,du\tag{4.56a}$$

$$= n\left(\frac{m}{2\pi kT}\right)^{\frac{1}{2}}\left(-\frac{kT}{m}\right)\exp\left[-\frac{mu^2}{2kT}\right]_0^\infty\tag{4.56b}$$

$$= n\left(\frac{kT}{2\pi m}\right)^{\frac{1}{2}} = \frac{n}{4}\left(\frac{8}{\pi}\times\frac{kT}{m}\right)^{\frac{1}{2}}.\tag{4.56c}$$

Comparison with equation (4.54) shows that this is equal to $\frac{1}{4}n\bar{c}$.

### 4.4.4 *Maxwell's derivation of the velocity distribution in a gas*

The velocity distribution derived above makes use of the Boltzmann distribution. It is interesting to see how Maxwell derived the velocity distribution before the more general ideas of Boltzmann had been developed. The following is based on Maxwell's paper published in 1860.

Let $n$ be the total number of molecules, and let $u$, $v$, $w$, be the components of the velocity of each molecule in three rectangular directions. Then if the number of molecules for which velocity component $u$ lies between $u$ and $u + du$ is $nf(u) \, du$, where $f(u)$ is the function to be determined, the number of molecules for which $v$ lies between $v$ and $v + dv$ will be $nf(v) \, dv$ and similarly for $w$, where $f$ always stands for the same function since there is no preferred direction in the gas.

Now the existence of the velocity $u$ does not in any way affect that of the velocities $v$ and $w$ since these are all at right angles to each other and independent. Consequently the number of molecules whose velocity components lie simultaneously between $u$ and $u + du$, between $v$ and $v + dv$ and between $w$ and $w + dw$ is

$$dn = nf(u)f(v)f(w) \, du \, dv \, dw. \tag{4.57}$$

Let this now refer to all those molecules which have a resultant velocity $c$, where

$$c^2 = u^2 + v^2 + w^2. \tag{4.58}$$

We now consider the condition under which the components $u$, $v$, $w$, can vary whilst $c$ remains constant. This is found by differentiating equation (4.58) and putting $dc = 0$; we obtain

$$u \, du + v \, dv + w \, dw = 0. \tag{4.59}$$

Since no direction is preferred over any other, it follows that $dn$ in equation (4.57) must remain constant whatever the individual values of $u$, $v$ and $w$, provided these satisfy equation (4.59). This implies that

$$\frac{d}{du}(dn) \, du + \frac{d}{dv}(dn) \, dv + \frac{d}{dw}(dn) \, dw = 0. \tag{4.60}$$

From equation (4.57) this gives

$$f'(u) \, du \, f(v)f(w) + f'(v) \, dv \, f(u)f(w) + f'(w) \, dw \, f(u)f(v) = 0,$$

where $f'(u)$ is the differential of $f(u)$ with respect to $u$, etc. Dividing throughout by $f(u) \, f(v) \, f(w)$ we obtain

$$\frac{f'(u)}{f(u)} \, du + \frac{f'(v)}{f(v)} \, dv + \frac{f'(w)}{f(w)} \, dw = 0. \tag{4.61}$$

In order to solve equations (4.59) and (4.61), we multiply equation (4.59) by an arbitrary constant $\lambda$ and add it to equation (4.61). The result is

$$\left(\frac{f'(u)}{f(u)} + \lambda u\right) du + \left(\frac{f'(v)}{f(v)} + \lambda v\right) dv + \left(\frac{f'(w)}{f(w)} + \lambda w\right) dw = 0.$$

Each term must identically be equal to zero, and since $du, dv, dw$, although very small are not themselves zero, the quantities in the brackets must be zero. Hence

$$\frac{f'(u)}{f(u)} + \lambda u = 0 \quad \text{or} \quad \frac{f'(u)}{f(u)} = -\lambda u.$$

Integrating this relation we have

$$\ln f(u) = -\frac{\lambda u^2}{2} + A,$$

where $A$ is an integration constant. This may be written in the form

$$f(u) = B \exp\left(-\frac{\lambda u^2}{2}\right).$$

As Maxwell remarks, if $\lambda$ were negative, the number of molecules would be infinite, so consequently, $\lambda$ must be positive. The number of molecules with velocities between $u$ and $u + du$ then becomes

$$dn = nf(u)\,du = nB \exp\left(-\frac{\lambda u^2}{2}\right).$$

There is no way of establishing the value of $\lambda$ except by calculating the mean square velocity $\overline{c^2}$ and making use of the gas equation: $\frac{1}{3}m\overline{c^2} = kT$.

Figure 32. (*a*) Method due to Zartman and Ko for determining the velocity distribution of bismuth atoms: the atoms are evaporated, collimated by the slits $S_1$, $S_2$, $S_3$ and enter the rotating drum. They fall on the cold plate P where they condense. (*b*) Results obtained in a later experiment by Esterman, Simpson and Stern.

(a)

(b)

density

velocity $c$

× × ×  experimental
——  theoretical

It is then found that $\lambda$ has the value $m/kT$. The result is

$$dn = nB \exp\left(-\frac{mu^2}{2kT}\right) du,$$

which is identical with equation (4.42).

### 4.4.5 *Experimental determination of velocity distribution*

In most of the earlier experimental studies of the velocity distribution, a furnace was used to produce a vapour of metal atoms, which behave like a monatomic gas, at the temperature of the enclosure. They were allowed to condense on a cold surface and, by the ingenious use of moving shutters, atoms with different velocities were caught at different points on the surface. The intensity of condensed atoms provides then a measure of the relative number of atoms within that velocity range. One method due to I. F. Zartman (1931) and C. C. Ko (1934) is illustrated in figure 32($a$). Atoms of bismuth are produced by a furnace and the vapour is collimated by a series of slits, $S_1$, $S_2$, $S_3$; the vapour beam reaches a drum rotating at a speed of 6000 rev. per min. and can only enter the drum at the slit S. The atoms then strike the plate P where they condense; the fastest ones reaching A, and slower ones B and the slowest C.

A more elegant and refined experiment by I. Esterman, O. C. Simpson and O. Stern (1947) gave results shown in figure 32($b$). The agreement with the Boltzmann distribution is surprisingly good.

## 4.5 **Thermal fluctuations in a gas**

We discuss here the fluctuations in density of a gas resulting from the wide range of molecular velocities, mean free paths, etc. Instead of tackling this analytically in terms of these parameters we shall proceed in the following way. We consider what would happen if a small change in volume and pressure occurred in a gas: we calculate the stored energy; we find that this is proportional to the square of the volume change. Consequently if such a change is due to thermal fluctuations the Boltzmann principle can be applied and the average value of the stored energy can be put equal to $\frac{1}{2}kT$.

We start with $N$ molecules of a gas in a cylinder at volume $V$, with a piston exerting an equilibrium pressure $P$. Then

$$PV = NkT. \tag{4.62}$$

If a small pressure $p$ is added externally to the piston the volume of the gas will diminish by $v$. Under overall isothermal conditions we have

$$(P+p)(V-v) = PV.$$

Hence

$$p = \frac{Pv}{V-v} \approx \frac{Pv}{V}. \tag{4.63}$$

If $p$ is increased to produce a further incremental change $\mathrm{d}v$, the work done on the gas is

$$p\,\mathrm{d}v = \frac{P}{V} v\,\mathrm{d}v. \tag{4.64}$$

In increasing $p$ from zero to its final value $p_0$ producing a final volume change $v_0$ the work done on the gas is

$$w = \int_0^{v_0} p\,\mathrm{d}v = \frac{P}{2V} v_0^2. \tag{4.65}$$

It is seen that work is done on the gas whether $v_0$ is positive or negative. The gas behaves as a 'spring', i.e. as an elastic medium of bulk modulus $P$.

We may forget about the piston and cylinder and accept equation (4.65) as a description of the extra energy acquired by a volume of gas $V$, nominal pressure $P$, if it undergoes a volume change $v_0$. We now consider the equilibrium between a large number of such volumes, separated from one another by a perfectly flexible, thermally conducting envelope. Then as a result of random molecular motions the volume of one envelope will be somewhat greater, another somewhat less than the equilibrium volume $V$. The fluctuating energy of each envelope is given by equation (4.65). We can at once apply the Boltzmann concept by noting that this relation gives an energy proportional to the square of a parameter $(v_0)$. The envelope behaves as though its variation in volume corresponds to one degree of freedom. Consequently if $\overline{v_0^2}$ is the mean square fluctuation in volume we have

$$\frac{P}{2V} \overline{v_0^2} = \tfrac{1}{2}kT. \tag{4.66}$$

Replacing $P$ by $NkT/V$ from equation (4.62) we obtain

$$\frac{\overline{v_0^2}}{V^2} = \frac{1}{N} \tag{4.67}$$

or

$$(v_0)_{\text{r.m.s.}} = V/N^{\frac{1}{2}}.$$

The volume fluctuates between $V + (v_0)_{\text{r.m.s.}}$ and $V - (v_0)_{\text{r.m.s.}}$, i.e. between $V(1 + 1/N^{\frac{1}{2}})$ and $V(1 - 1/N^{\frac{1}{2}})$. Consequently the number of molecules per unit volume (the 'number density') undergoes a fractional fluctuation lying between

$(1 - 1/N^{\frac{1}{2}})$ and $(1 + 1/N^{\frac{1}{2}})$.                                    (4.68)

At s.t.p. $N = 3 \times 10^{28}$ molecules per $m^3$. The fraction $1/N^{\frac{1}{2}}$ is only $6 \times 10^{-15}$. Such a fractional fluctuation would be quite undetectable. However, if one considers a very small volume containing on average, say, 100 molecules, the number will fluctuate between $100(1 - \frac{1}{10})$ and $100(1 + \frac{1}{10})$, i.e. between 90 and 110.

### 4.6 Thermal energy of molecules

#### 4.6.1 *Specific heat capacities, number of degrees of freedom*

We have already seen that for a perfect gas in thermal equilibrium each molecule possesses an average thermal energy of translation of amount $\frac{3}{2}kT$. Since a molecule has 3 independent degrees of translational freedom we may associate energy $\frac{1}{2}kT$ with each degree of freedom. This may be generalized to apply to all particles in thermal equilibrium; however, for macroscopic bodies the resulting velocity is rather small. For example the mean thermal velocity of a cricket ball in equilibrium with its surroundings at room temperature (say 300 K) is about $10^{-10}$ m s$^{-1}$.

The result may be generalized even further. If we define the number of degrees of freedom as the number of independent squared terms that enter into the total energy of a particle, then each has associated with it an average thermal energy of amount $\frac{1}{2}kT$. For example the kinetic energy due to translation is determined solely by (linear velocity)$^2$ in each of these independent coordinates; each degree of freedom has thermal energy $\frac{1}{2}kT$. The energy of rotation of a body about a specified axis is determined by (angular velocity)$^2$, so that each of these degrees of freedom has energy $\frac{1}{2}kT$. Vibrations present a special problem. If the equation of motion, for a body of mass $m$, is

$$\ddot{x} = -\omega^2 x, \quad \text{with solution } x = A \sin \omega t,                                    (4.69)$$

the kinetic energy at any instant is $\frac{1}{2}m\dot{x}^2$, whilst the potential energy is easily shown to be equal to $\frac{1}{2}m\omega^2 x^2$. By substituting in equation (4.69) we find that the sum of kinetic and potential energy is $\frac{1}{2}mA^2\omega^2$, so it is independent of $x$ and $\dot{x}$. Nevertheless at any instant the kinetic energy is proportional to $\dot{x}^2$ and the potential energy to $x^2$. Each of these contributes $\frac{1}{2}kT$ to the average thermal energy.

Summarizing we may write: thermal energy per degree of translational freedom $= \frac{1}{2}kT$, per degree of rotational freedom $= \frac{1}{2}kT$ and per vibrational mode (equivalent to two degrees of freedom*) $= kT$.

*Some authorities refer to vibration as possessing a *single* degree of freedom of energy $kT$.

### 4.6.2 *Internal energy of a gas*

If we consider the energy which arises solely from thermal motion (not excited electronic states), the internal energy $u$ per molecule of a gas may be deduced for monatomic, diatomic and polyatomic molecules.

Monatomic gas: only 3 translations $u = \frac{3}{2}kT$ per molecule.

Diatomic gas: 3 translations
   2 rotations

$u = \frac{5}{2}kT$ per molecule plus any vibrational energy.

The third rotation about the axis of the molecule does not occur for reasons to be described later.

Polyatomic gas: 3 translations
   3 rotations (if non-linear)

$u = \frac{6}{2}kT$ plus any vibrational energy.

These results are summarized in Table 15.

Table 15.  *Internal energy U and specific heat capacities C of gases*

| Gas | Monatomic | Diatomic | Polyatomic |
|---|---|---|---|
| $f_{tr}$ | 3 | 3 | 3 |
| $f_{rot}$ | 0 | 2 | 3 |
| $f_{vib}$ | 0 | assumed 0 | assumed 0 |
| Total | 3 | 5 | 6 |
| $u$ mol$^{-1}$ | $\frac{3}{2}kT$ | $\frac{5}{2}kT$ | $\frac{6}{2}kT$ |
| $U$ mol$^{-1}$ | $\frac{3}{2}RT$ | $\frac{5}{2}RT$ | $\frac{6}{2}RT$ |
| $C_V = \left(\dfrac{\partial U}{\partial T}\right)_V$ | $\frac{3}{2}R$ | $\frac{5}{2}R$ | $3R$ |
| $C_P = C_V + R$ | $\frac{5}{2}R$ | $\frac{7}{2}R$ | $4R$ |
| $\gamma = \dfrac{C_P}{C_V}$ | $\frac{5}{3}$ | $\frac{7}{5}$ | $\frac{4}{3}$ |
| $C_V$ in units of $R/2$ mol$^{-1}$: | | | |
| Theoretical | 3 | 5 | 6 |
| Observed | Ar:3 | $H_2$:4.9 | $H_2S$:6 |
|  | He:3 | $O_2$:5.0 | $CS_2$:10 |
|  |  | $N_2$:4.6 |  |
|  |  | $Cl_2$:6 |  |

For monatomic gases the agreement is excellent. For diatomic gases the agreement is good, but the value given for chlorine seems to indicate that it possesses additional vibrational energy at room temperature. With polyatomic molecules the behaviour is very complicated but it would seem that, in general $C_V > 6 \times R/2$ mol$^{-1}$.

In general, if one knows the number of degrees of freedom $f$ which are 'active' one can write

$$U = f \times \frac{RT}{2},$$

so that

$$C_V = \tfrac{1}{2}Rf$$

and

$$\gamma = \frac{C_P}{C_V} = 1 + \frac{2}{f}. \tag{4.70}$$

### 4.6.3 *Gas kinetic theory of adiabatic expansion*

Before we go on to discuss in greater detail the vibrational and rotational energy of a molecule, we may at once use the ideas expressed in the previous section to deduce the adiabatic gas equation for monatomic and polyatomic gases.

In an adiabatic expansion all the energy is taken from the internal energy of the gas. We first consider a monatomic gas where the internal energy arises solely from the kinetic energy of translation.

From the kinetic theory we have:

$$PV = \tfrac{2}{3} \text{ (total kinetic energy } Z )$$

$$Z = \tfrac{3}{2} PV = \text{internal energy } U.$$

The work done by the gas in an adiabatic expansion is

$$P\,dV = -dU = -dZ = -d(\tfrac{3}{2}PV).$$

Therefore

$$P\,dV = -\tfrac{3}{2}P\,dV - \tfrac{3}{2}V\,dP$$

or

$$\tfrac{5}{2}P\,dV + \tfrac{3}{2}V\,dP = 0$$

or

$$5\frac{dV}{V} + 3\frac{dP}{P} = 0.$$

Hence $\underline{PV^{\frac{5}{3}} = constant.}$ \hfill (4.71)

For a diatomic gas possessing two rotational degrees of freedom, the internal energy is

$$U = \text{kinetic energy} + \text{rotational energy}$$
$$= \tfrac{3}{2}PV + RT = \tfrac{3}{2}PV + PV$$
$$= \tfrac{5}{2}PV.$$

For an adiabatic expansion $P\,dV = -d(\tfrac{5}{2}PV)$. This leads, as above to

$$PV^{7/5} = \text{constant.} \tag{4.72}$$

### 4.6.4 *Rotational and vibrational degrees of freedom*

At room temperature the specific heat capacity $C_V$ of hydrogen is about $5 \times R/2 \text{ mol}^{-1}$. As the temperature is lowered $C_V$ decreases until at $50 K$, $C_V$ has a value of only about $3 \times R/2 \text{ mol}^{-1}$. It is apparent that the rotational degrees of freedom have gradually disappeared and that at about 50 K the molecule has only translational energy. The reason for this is that, according to the quantum theory, rotors can possess only discrete energy levels. For a symmetrical diatomic molecule rotating about an axis, for which the moment of inertia is $I$, the possible energy levels are

$$E_{\text{rot}} = n(n+1)\frac{h^2}{8\pi^2 I}, \tag{4.73}$$

where $h$ is Planck's constant $(6.6 \times 10^{-34} \text{ J s}^{-1})$ and $n$ is an integer. The smallest value of rotational energy permitted occurs when $n = 1$ and has the value

$$(E_{\text{rot}})_{\min} = \frac{h^2}{4\pi^2 I}. \tag{4.74}$$

For a hydrogen molecule, $I$ about an axis normal to the bond is approximately $4.7 \times 10^{-48} \text{ kg m}^2$. Therefore

$$(E_{\text{rot}})_{\min} = 2 \times 10^{-21} \text{ J.}$$

But the translational energy $E_{\text{tr}} = \tfrac{3}{2}kT = 2 \times 10^{-23} \, T \text{ J}$, so that at 50 K

$$E_{\text{tr}} = 10^{-21} \text{ J.} \tag{4.75}$$

Therefore the translational energy is appreciably less than the lowest rotational energy level. By Boltzmann's principle the fraction of molecules capable of acquiring rotational energy $2 \times 10^{-21}$ J at 50 K will be

$$\exp\left(-\frac{\Delta\varepsilon}{kT}\right) \approx \exp\left(-3\right) \approx \tfrac{1}{20}. \tag{4.76}$$

Consequently below 50 K very few of the molecules will take up rotational motion. At room temperature (300 K), $kT$ will be large compared with the first rotational energy level, so a large fraction of the molecules will be able to take up the first and higher rotational energy levels.

Equation (4.74) also shows why rotating about the axis itself does not occur. The moment of inertia $I$ is so small that the lowest rotational energy state, $h^2/4\pi^2 I$, is enormous and can never be reached before thermal dissociation occurs.

Similar considerations apply to the vibrational degrees of freedom. If the vibrational frequency is $\nu$, the first energy state (ignoring zero-point energy) demands energy of amount equal to $h\nu$. For the hydrogen molecule $\nu \approx 2.6 \times 10^{14} \, s^{-1}$, so that $h\nu$ is about $2 \times 10^{-19} \, J$. For an appreciable fraction of molecules to be excited to this level $kT$ must have about the same value, so that $T$ must exceed 10 000 K. The molecule dissociates long before this. By contrast the chlorine molecule, which has a much lower natural frequency $\nu$, is able to take up appreciable vibrational energy at temperatures above 600 K so that its specific heat capacity approaches $C_V = 7 \times R/2 \, mol^{-1}$ at elevated temperatures (see figure 33).

### 4.6.5 *Calculation of vibrational energy as a function of temperature*

We reproduce here the standard treatment for a simple oscillator of natural frequency $\nu$. According to quantum theory the possible energy

Figure 33. Variation of specific heat capacity with temperature for $H_2$ and for $Cl_2$.

levels are

$$E_{\text{vib}} = (n + \tfrac{1}{2})h\nu, \tag{4.77}$$

where $n$ is an integer.

The zero-point energy $\tfrac{1}{2}h\nu$ is the vibrational energy the molecule possesses even at absolute zero, but since this is a constant we may ignore it at this stage of the computation and add it at the end. We put

$$E_{\text{vib}} = nh\nu. \tag{4.78}$$

Applying the Boltzmann function, the number of states with quantum number $n$ is

$$N_n = A \exp -\frac{nh\nu}{kT}$$

$$= A \exp(-n\beta), \quad \text{where} \quad \beta = \frac{h\nu}{kT}. \tag{4.79}$$

The total number $N$ of atoms is the sum of all $N_n$ from $n = 0$ to $n = \infty$,

$$N = A[1 + \exp(-\beta) + \exp(-2\beta) + \exp(-3\beta) + \cdots]$$

$$= A[1 + \exp(-\beta) + (\exp(-\beta))^2 + (\exp(-\beta))^3 + \cdots]$$

$$= \frac{A}{1 - \exp(-\beta)}. \tag{4.80}$$

This gives $A$ so that $N_n$ becomes

$$N_n = N[1 - \exp(-\beta)][\exp(-n\beta)]. \tag{4.81}$$

The total energy is the sum of $N_n E_{\text{vib}}$ where $E_{\text{vib}} = nh\nu$.

$$U = N[1 - \exp(-\beta)]$$
$$\times [\exp(-n\beta)] nh\nu \text{ summed for all values of } n$$
$$= N[1 - \exp(-\beta)]$$
$$\times \{h\nu[\exp(-\beta)] + 2h\nu[\exp(-\beta)]^2 + 3h\nu[\exp(-\beta)]^3 + \cdots\}$$
$$= N[1 - \exp(-\beta)]h\nu[\exp(-\beta)]\{1 + 2[\exp(-\beta)]$$
$$+ 3[\exp(-\beta)]^2 + \cdots\}$$
$$= N[1 - \exp(-\beta)]h\nu[\exp(-\beta)] \frac{1}{[1 - \exp(-\beta)]^2}$$
$$= Nh\nu \frac{[\exp(-\beta)]}{[1 - \exp(-\beta)]} = \frac{Nh\nu}{\exp(\beta) - 1}. \tag{4.82}$$

If we include the zero-point energy $\tfrac{1}{2}h\nu$ we have

$$U = Nh\nu\{\tfrac{1}{2} + [\exp(\beta) - 1]^{-1}\}. \tag{4.83}$$

At high temperatures $\exp \beta = \exp (h\nu/kT) \approx 1 + h\nu/kT$ so that

$$U \approx Nh\nu \left(\frac{1}{2} + \frac{kT}{h\nu}\right),$$

$$U = \tfrac{1}{2}Nh\nu + NkT. \tag{4.84}$$

Thus apart from zero-point energy each vibration carries average energy $kT$, so that the specific heat capacity contribution is $Nk = R$ for each vibration. At lower temperatures $[\exp (h\nu/kT) - 1]^{-1}$ and therefore the specific heat capacity falls off and tends to zero when $T = 0$ (see figure 34). This decrease in vibrational energy with temperature reproduces essentially the right-hand features of the specific heat capacity curves of hydrogen and chlorine in figure 33. We may also note in passing that this concept was applied by A. Einstein to explain the specific heat capacity of solids. He suggested that each atom has 3 independent vibrational modes of constant frequency $\nu$. The specific heat capacity (in this case it assumes no volume change and is therefore $C_V$) is $3R$ at high temperatures and falls to zero at low temperatures. We discuss this in greater detail in a later chapter.

### 4.6.6 *Rotational energy*

For a symmetrical diatomic molecule rotating about an axis for which the moment of inertia is $I$, the possible energy levels $E_{rot}$ are as given in equation (4.73). The application of the Boltzmann distribution is

Figure 34. Variation of vibrational specific heat capacity with temperature for a simple oscillation of natural frequency $\nu$. Specific heat capacity is in units of $R$ mol$^{-1}$, temperature in units of $k/h\nu$.

complicated because each state of quantum number $n$ consists of $g$ states of practically identical energy. The degeneracy factor $g$ has the value

$$g = 2n + 1. \tag{4.85}$$

One can build up the Boltzmann relation as before where the basic exponential term is

$$\exp\left(-\frac{E_{rot}}{kT}\right) \quad \text{or} \quad \exp\left(-n(n+1)\frac{h^2}{8\pi^2 I}\frac{1}{kT}\right).$$

The final integral has to be summed numerically. For high temperatures the energy for $N$ rotors becomes $NkT$. This is the classical result for a rotor with two degrees of freedom. In general the rotational energy falls off from $NkT$ when

$$kT < \frac{h^2}{8\pi^2 I}. \tag{4.86}$$

### 4.6.7 *Translational energy*

For a gas confined within a container, quantization of translational energy does occur, but the number of energy states is enormous and the spacing between energy levels is minute. We may therefore consider all energy states to be available as a continuous distribution, so that the classical Boltzmann distribution applies. Consequently each degree of translational freedom involves energy $\frac{1}{2}kT$ per molecule under all practical conditions.

### 4.6.8 *Equipartition as a broader principle*

In all that we have so far discussed we have considered systems for which the energy is a function of squared terms: each of these acquires an average energy of $\frac{1}{2}kT$. What of the other energy functions? Dr C. Hall has pointed out the following interesting point.

Suppose the energy is proportional to $x^m$, say $E = \alpha x^m$, then the average energy is given by

$$\bar{E} = \frac{\int_0^\infty \alpha x^m \exp(-\alpha x^m/kT)\,dx}{\int_0^\infty \exp(-\alpha x^m/kT)\,dx}. \tag{4.87}$$

This may be solved as follows. Let us write

$$y^m = \alpha x^m/kT \tag{4.88}$$

so that

$$dy = (\alpha/kT)^{1/m}\,dx.$$

Then equation (4.87) becomes

$$\bar{E} = \frac{kT \int_0^\infty y^m \exp(-y^m) \, dy}{\int_0^\infty \exp(-y^m) \, dy} = kT\left(\frac{I_N}{I_D}\right). \tag{4.89}$$

The denominator $I_D$ may be integrated by parts

$$I_D = \int_0^\infty \exp(-y^m) \cdot 1 \cdot dy$$

$$= [\exp(-y^m)y]_0^\infty + \int_0^\infty my^{m-1} \exp(-y^m) y \, dy. \tag{4.90}$$

The first term is zero. This leaves only the second term. Hence

$$I_D = m \int_0^\infty y^m \exp(-y^m) \, dy. \tag{4.91}$$

We note that the integral in equation (4.91) corresponds exactly to the integral $I_N$ in the numerator of equation (4.89). Thus

$$I_D = mI_N.$$

Consequently

$$\bar{E} = kT\frac{I_N}{I_D} = \frac{1}{m} kT. \tag{4.92}$$

Typical examples are:

$E = \alpha x, \bar{E} = kT$;

> this corresponds to a particle in a uniform gravitational field.

$E = \alpha x^2, \bar{E} = \frac{1}{2}kT$;

> this corresponds to the classical definition of a degree of freedom.

$E = \alpha x^3, \bar{E} = \frac{1}{3}kT$;

$E = \alpha x^4, E = \frac{1}{4}kT$;

> these do not correspond to common physical phenomena.

### 4.6.9 *The basic form of the Boltzmann distribution*

This section is included for completeness and may be omitted by the general reader.

In the preceding sections we have provided a rather makeshift version of the Boltzmann distribution. The full form relies on a very subtle theorem due to Liouville. We first consider an assembly of particles, atoms or molecules in equilibrium and specify that a given particle has momentum lying between $p_x$ and $(p_x + dp_x)$, $p_y$ and $(p_y + dp_y)$, $p_z$ and

$(p_z + dp_z)$ and has coordinates lying between $x$ and $(x + dx)$, $y$ and $(y + dy)$, $z$ and $(z + dz)$. We then consider how a given particle (or number of particles) will enter and leave this volume element $dp_x \, dp_y \, dp_z \, dx \, dy \, dz$ of 'phase space'. Analysis shows that, for particles obeying the laws of classical mechanics, if a given particle occupies an element of phase space at a given time it will occupy another element of equal volume at a later time. This implies that all phase space elements of the same volume have the same probability of being occupied by a given particle. This is Liouville's theorem.

We may now divide phase space into a number of elementary cells of equal volume, i.e. equal *a priori* probability of occupancy, and consider how we could distribute all the particles of the system in these cells to give a constant (specified) total energy of the system and a maximum total probability, i.e. the one that in fact will occur. Boltzmann showed that the probability of finding a particle with a given momentum at a given place is proportional to $\exp(-E/kT)$ per unit elementary volume of phase space, where $E$ is the *total* energy of the particle. Thus

$$\text{probability} = A \, \exp(-E/kT) \, dp_z \, dp_y \, dp_z \, dx \, dy \, dz. \qquad (4.93)$$

↓        ↓

volume   volume

of       of

momentum real

space    space

For a particle with kinetic energy $K$ (function of velocity, i.e. of momentum) and potential energy $U$ (function of position) the probability is given by equation (4.93) with $E = K + U$. This can be written as

$$A \, \exp(-K/kT) \, dp_x \, dp_y \, dp_z \times \exp(-U/kT) \, dx \, dy \, dz. \qquad (4.94)$$

↓                ↓

Probability of having   probability of being

a given velocity        at a given place

This is therefore equivalent to the product of two independent probabilities. The splitting-up of the probability into the product of a number of independent terms is one of the characteristic features that follows from the exponential function in the Boltzmann distribution. Many of the calculations in the previous sections are clear examples of this. If no potential energy is involved (as in a perfect gas) we are left with the first term which provides the basis for the Maxwell velocity distribution. Noting that $dp_x = m \, du$, etc., we obtain for the probability

$$A \exp(-E/kT) \, du \, dv \, dw. \qquad (4.95)$$

This is *not* the same as

$$A \exp(-E/kT) \, dE. \qquad (4.96)$$

Indeed, it is generally incorrect to write the probability as $\exp(-E/kT) \, dE$. There is only one simple case where it does apply, i.e. to a two-dimensional gas where we can replace $u^2 + v^2$ by, say, $c_1^2$ and take an annular element of velocity space $2\pi c_1 \, dc_1$. This element then corresponds to $(2\pi/m) \, dE$.

The Boltzmann distribution has a quantum form which we have already used. Under quantum conditions we do not refer to values of momentum or position but describe the system as being in a quantum state with a probability proportional to $\exp(-E/kT)$.

## 4.7 Macroscopic examples of equipartition of energy

We conclude this chapter with four examples of thermal motion occurring in systems other than gases.

### 4.7.1 *Galvanometer mirror*

A delicately suspended galvanometer mirror can be seen to undergo small random oscillations which are due to its thermal energy. The mirror has a single degree of oscillation about its axis so that we attribute to its mean square velocity $(\overline{\omega^2})$, energy $\frac{1}{2}kT$, and to its mean square angular deflection $(\overline{\theta^2})$, energy $\frac{1}{2}kT$. We may write

$$\tfrac{1}{2}kT = \tfrac{1}{2}I(\overline{\omega^2}) = \tfrac{1}{2}\tau(\overline{\theta^2}), \qquad (4.97)$$

where $I$ is the moment of inertia, $\tau$ is the torsion constant. If we determine the torsional stiffness of the suspension we may then calculate $(\overline{\theta^2})$. This tells us the amount by which any galvanometer reading will fluctuate as a result of thermal motion. These fluctuations do not depend on the presence of air surrounding the mirrors; it is true that gas molecules will buffet the mirror and contribute to the oscillation, but they will also contribute to the damping, so the net effect is unchanged. The mirror behaves like a large molecule with a single mode of torsional oscillation. In a perfect vacuum the oscillations can be considered as arising from the random absorption and radiation of electromagnetic energy associated with the temperature $T$. The resulting behaviour is the same.

### 4.7.2 *Sedimentation*

As we saw earlier, particles in thermal equilibrium behave like gas molecules of large molecular weight. If they are in an evacuated container the sedimentation due to gravity is so rapid that they virtually all lie at the bottom of the container. If, however, they are suspended in a liquid of very nearly the same density the effective gravitational field is greatly reduced and there may be an appreciable variation in particle concentration with distance from the bottom of the container, as given by equation (4.35).

### 4.7.3 *Electrical noise*

The conduction electrons in a metal may be regarded as a gas with random velocities. These give rise to electrical 'noise'. The r.m.s. potential fluctuation across a resistor of resistance $\Omega$ is given by

$$\text{r.m.s. potential} = [4\Omega k T (f_2 - f_1)]^{\frac{1}{2}}, \tag{4.98}$$

where $f_2 - f_1$ is the bandwidth frequency over which the measurements are made. This result at once shows that there is a limit to useful amplification. If the original signal is too small (e.g. comparable with the random electrical noise) no amount of amplification will improve reception.

Similarly in the thermionic emission of current $I_0$ it may be shown that there is a fluctuation $i$ in the emitted current given by

$$|\overline{i^2}|^{\frac{1}{2}} = \left(\frac{I_0 e}{\tau}\right)^{\frac{1}{2}}, \tag{4.99}$$

where $e$ is the electronic charge and $\tau$ the time constant of the measuring system. This effect has indeed been used as an independent means of determining $e$; the result agrees with other methods to within 1 per cent.

### 4.7.4 *Brownian motion*

The English botanist R. Brown observed in 1827 that finely divided particles in suspension undergo ceaseless movement. This motion was analysed by Einstein in 1906; the following derivation is based on a treatment developed by P. Langevin in 1908. We first assume that the particles possess thermal kinetic energy in one dimension of amount $\frac{1}{2}kT$.

If at any instance the molecules of the liquid collide with an individual particle to produce a net force $X$ in the $x$ direction, we may write the equation of motion of the particle in the form

$$m\ddot{x} + 6\pi a \eta \dot{x} = X, \tag{4.100}$$

where $m$ is the mass, $a$ the radius of the particle (assumed spherical) and $\eta$ the viscosity of the fluid. In this equation the first term represents the inertial force on the particle and the second the frictional force due to the viscous resistance of the fluid; this is Stokes' well-known relation for viscous forces on a sphere. The third term represents the force due to molecular collisions. Thus from the point of view of collisions we treat the fluid as a collection of individual molecules, but from the point of view of the resistive forces we treat the fluid as a continuum. Finally, this equation refers only to motions of the particle produced by collisions; the intrinsic motion of the particle by virtue of its own kinetic energy ($\frac{1}{2}kT$) emerges from the analysis.

We multiply both sides of equation (4.100) by $x$ and then average over a large number of collisions. Then for any fixed specified value of $x$, for every value of $X$ there is an equal chance of $-X$ occurring so that $\Sigma xX = 0$.

$$\Sigma m\ddot{x}x + 6\pi a\eta \Sigma \dot{x}x = \Sigma xX = 0 \qquad (4.101)$$

or

$$m\overline{\ddot{x}x} + 6\pi a\eta \,\overline{\dot{x}x} = 0.$$

We note that $d(\dot{x}x)/dt = \ddot{x}x + \dot{x}^2 = \ddot{x}x + u^2$, where $u$ is the velocity in the $x$ direction. Then

$$m\frac{d}{dt}(\overline{\dot{x}x}) - m\overline{u^2} + 6\pi a\eta \,\overline{\dot{x}x} = 0. \qquad (4.102)$$

It turns out that the first term implies a transient at the beginning of the collision which has a negligible effect on the subsequent displacement.[†]

Hence

$$6\pi a\eta \,\overline{\dot{x}x} = m\overline{u^2} = kT, \qquad (4.103)$$

or

$$3\pi a\eta \frac{d}{dt}(\overline{x^2}) = kT. \qquad (4.104)$$

---

[†] More precisely we may write $\dot{x}x$ as $\frac{1}{2}dx^2/dt$ which we may call $z$. Equation (4.102) then becomes

$$m\dot{z} + 6\pi a\eta z = mu^2 = kT.$$

The solution is

$$z = \frac{kT}{6\pi a\eta} + A \exp\left(\frac{-6\pi a\eta t}{m}\right),$$

where $A$ is a constant of integration. The quantity $6\pi an/m$ is extremely large so that even for minute values of $t$ the exponential term becomes negligible. During the total time of the collision, $z$ is thus given essentially by the first term on the RHS of the equation. This is indeed the same equation as (4.104).

In a time interval $\tau$ the mean square displacement $\overline{\Delta x^2}$ is then

$$\overline{\Delta x^2} = \frac{kT\tau}{3\pi a\eta} = \frac{RT\tau}{3\pi a\eta N_A}, \tag{4.105}$$

where $N_A$ is Avogadro's number. A given particle is observed say every 30 seconds and the displacement $\Delta x$ in the $x$ direction is observed for each interval of time. These values are squared and the mean value is calculated; this gives $\overline{\Delta x^2}$. This is an average path for the time interval $\tau = 30$ seconds. The value of $\overline{\Delta x^2}$ so found may be compared with the right-hand term of equation (4.105). Experiments show that the results agree very well with calculation.

There are two points of interest. First, the displacement does not depend on the mass of the particle (only on its radius). Secondly, although the individual particle has intrinsic motion, which has nothing to do with the surrounding liquid, the distance it moves about an equilibrium position *is* determined by the viscous damping of the liquid. In a descriptive way we may say that a viscous force $6\pi a\eta \dot{x}$ acting over a distance $x$ consumes the thermal energy $\frac{1}{2}mu^2$. The particle then reacquires its thermal energy and the process is repeated. Apart from a numerical factor this is essentially the result implied by equation (4.103).

If then equation (4.103) is a description of the way in which viscous damping consumes thermal energy, is it not possible to apply it to the behaviour of a gas molecule within its own gas? We have already derived a relation for the viscosity of a gas, namely $\eta = \frac{1}{3}mn\bar{c}\lambda$. However, this relation can only be used when the distance between the moving bodies is larger than the mean free path $\lambda$. On the other hand, in the derivation of Stokes' law for the viscous resistance experienced by a particle moving through a gas, the particle must be large compared with the mean free path of the gas so that it experiences the buffeting by the gas molecules as though the gas were a continuum. Thus we cannot really use Stokes' law if the sphere is the gas molecule itself. However, as an extreme case let us see what happens if we take a heavily condensed gas where $\lambda = a$. We have

$$\eta = \tfrac{1}{3}mn\bar{c}a; \dot{x} \approx u; x \approx a.$$

Then the LHS of equation (4.103) becomes

$$6\pi a\eta \overline{\dot{x}x} = 6\pi a\tfrac{1}{3}mn\bar{c}aua = 2\pi a^3 nm\bar{c}u.$$

But

$$\bar{c} \approx \sqrt{3}(u).$$

Hence

$$6\pi a n \overline{\dot{x}\dot{x}} \approx 10a^3 nmu^2. \tag{4.106}$$

But each sphere of radius $a$ occupies a cube of volume $8a^3$, so that $8a^3 n = 1$. The value of equation (4.106) is thus approximately $mu^2$, which agrees with the RHS of equation (4.103). This is, of course, pushing the model to an extreme, but it shows again that the particle considered in the analysis of Brownian motion is essentially no different from a molecule, except that it is very much larger.

This has been a long chapter, and we now conclude it with a relevant quotation from Lucretius:

This process, as I might point out, is illustrated by an image of it that is continually taking place before our very eyes. Observe what happens when sunbeams are admitted into a building and shed light on its shadowy places. You will see a multitude of tiny particles mingling in a multitude of ways in the empty space within the light of the beam, as though contending in everlasting conflict, rushing into battle rank upon rank with never a moment's pause in a rapid sequence of unions and dis-unions. From this you may picture what it is for the atoms to be perpetually tossed about in the illimitable void. To some extent a small thing may afford an illustration and an imperfect image of great things. Besides, there is a further reason why you should give your mind to these particles that are seen dancing in a sunbeam: their dancing is an actual imitation of underlying movements of matter that are hidden from our sight. There you will see many particles under the impact of invisible blows changing their course and driven back upon their tracks, this way and that, in all directions. You must understand that they all derive this restlessness from the atom. It originates with the atoms, which move themselves. Then those small compound bodies that are least removed from the impetus of the atoms are set in motion by the impact of their invisible blows and in turn cannon against slightly larger bodies. So the movement mounts up from the atoms and gradually emerges to the level of our senses, so that those bodies are in motion that we see in sunbeams, moved by blows that remain invisible.

(It is interesting to note that almost two thousand years later Perrin used the identical argument in his studies of Brownian motion of gamboge particles in a liquid: he suggested that this could be projected optically before an audience to demonstrate the molecular movement within the liquid.)

Professor S. Sambursky, in his perceptive study *The Physical World of the Greeks*,† adds the comment that this remarkable description, 'perfectly describes and explains the Brownian movement by a wrong example. The movement of dust particles as seen by the naked eye in sunlight are caused by air-currents: the real phenomenon postulated by Lucretius

---

† S. Sambursky, *The Physical World of the Greeks* (Routledge, 1956).

on the basis of abstract reasoning can only be seen in a microscope.'
Further, as we have pointed out previously, the random motion of fine
particles is an intrinsic property of any assembly of free entities in thermal
equilibrium, whether they be gas molecules or solid particles. Nothing,
however, can detract from the beauty and elegance of this inferential type
of argument.

*Integrals*

Values of

$$\int_0^\infty x^n \, e^{-ax^2} \, dx:$$

$$\int_0^\infty e^{-ax^2} \, dx = \frac{1}{2}\left(\frac{\pi}{a}\right)^{\frac{1}{2}}$$

$$\int_0^\infty x \, e^{-ax^2} \, dx = \frac{1}{2a}$$

$$\int_0^\infty x^2 \, e^{-ax^2} \, dx = \frac{1}{4}\left(\frac{\pi}{a^3}\right)^{\frac{1}{2}}$$

$$\int_0^\infty x^3 \, e^{-ax^2} \, dx = \frac{1}{2a^2}$$

$$\int_0^\infty x^4 \, e^{-ax^2} \, dx = \frac{3}{8}\left(\frac{\pi}{a^5}\right)^{\frac{1}{2}}$$

In general

$$\int_{-\infty}^\infty x^n \, e^{-ax^2} \, dx = 2\int_0^\infty x^n \, e^{-ax^2} \, dx \text{ for } n \text{ even}$$

$$= 0 \text{ for } n \text{ odd}$$

# 5

## *Imperfect gases*

Many gases even at room temperature deviate markedly from the 'ideal' gas relation $PV = RT$. Similar deviations are observed with noble gases at higher pressures and lower temperatures. In this chapter we shall discuss the behaviour of such 'imperfect' gases and show that their properties can be explained in terms of the 'ideal' gas theory, modified to allow for the finite volume of the gas molecules and for the existence of intermolecular forces.

### 5.1 **Deviations from perfect gas behaviour**
#### 5.1.1 *The virial equation*
In general, instead of the relation $PV = $ constant at a fixed temperature, imperfect gases obey an empirical power equation of the form

$$PV = A + BP + CP^2 + DP^3. \tag{5.1}$$

This is known as a virial equation and $A$, $B$, $C$ and $D$ as virial coefficients. The most important coefficient, apart from $A$, is $B$, since $C$ and $D$ are usually very small. At low temperatures $B$ is negative, at higher temperatures positive and at some intermediate temperature $T_B$ it has the value zero. At this temperature, if the pressures are not too high we see at once that equation (5.1) reduces to

$$PV = A = \text{constant}, \tag{5.2}$$

i.e. Boyle's law becomes approximately true again. For this reason $T_B$ is called the Boyle temperature.

Plots of $PV$ against $P$ are shown schematically in figure 35. If we measure the slope of the curves near the $PV$ axis, i.e. near $P = 0$, we have

$$\frac{\partial(PV)}{\partial P} = B + 2CP + \cdots$$

$$= B \text{ for } P \to 0. \tag{5.3}$$

The slope near $P = 0$ will, therefore, be zero at the Boyle temperature. It is also easy to show that the curve connecting the minima is a parabola.

### 5.1.2 *Andrews' experiments on carbon dioxide*

T. Andrews (1813–85) carried out a series of important and classical experiments on the compressibility of gases and the conditions under which they can be liquefied. His first paper appeared in 1861 and one of his major surveys was given in his Bakerian lecture of 1869. We summarize his main conclusions as exemplified by the behaviour of $CO_2$ (see figure 36).

(*a*) Above a temperature of about 50 °C, the compressibility of $CO_2$ resembles that of a perfect gas.

(*b*) At, say, 21 °C compression produces liquefaction.

From B to C the material behaves as a gas, from C to D it condenses at a constant pressure, and at D it is completely liquid. From D to K the slope is very steep because a liquid is relatively incompressible.

Figure 35. Variation of $PV$ with $P$ for a real gas. At one special temperature, where the initial tangent is horizontal, $PV$ is constant in the lower pressure range; this is the Boyle temperature $T_B$.

(c) At 31.1 °C there is no liquefaction although the isotherm is very distorted from the ideal gas isotherm.

(d) At 30.9 °C liquefaction just occurs under compression.
Above this temperature, liquefaction cannot be produced. This is known as the critical temperature $T_c$, the temperature above which liquefaction cannot be achieved, however high the pressure. The pressure at which liquefaction just occurs is called the critical pressure $P_c$, and the corresponding volume (for one mole of the substance) is called the critical volume $V_c$.

### 5.1.3 *Continuity of liquid and gaseous state*

Let us consider two isothermals, one above $T_c$ and one below. Consider the points J and K on these isothermals at some pressure level above $P_c$. At J the substance is purely a gas, at K purely a liquid; this means that if we keep the pressure constant at J but steadily reduce the temperature we may pass to the liquid phase at K. This, in turn, implies that above $P_c$ it is possible, simply by cooling, to pass from the gaseous to the liquid state without any mixture of phases.

### 5.1.4 *Significance of* $T_c$

The critical temperature may be easily understood in terms of intermolecular forces. Suppose figure 37 represents the potential energy between one molecule and a single neighbour, as a function of separation.

Figure 36. Variation of pressure $P$ with volume $V$ for $CO_2$ (based on Andrews' results).

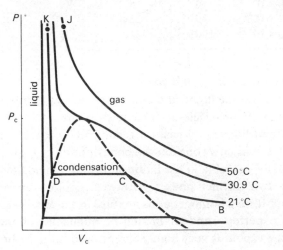

It is clear that if the thermal energy of the molecule is greater than $\Delta\varepsilon$, one molecule can always escape from its neighbours. However closely they are pressed together (within limits) the thermal energy will always be sufficient to overcome the attraction between the molecules.

Now the thermal energy of each molecule is of order $kT$, where $k$ is Boltzmann's constant and equals $1.4 \times 10^{-23}$ J K$^{-1}$ and $T$ is the absolute temperature. Thus a molecule can always escape from its neighbour if $kT > \Delta\varepsilon$ or

$$T > \frac{\Delta\varepsilon}{k}. \tag{5.4}$$

The substance is therefore unliquefiable for temperatures above

$$T_c = \frac{\Delta\varepsilon}{k}.$$

Some typical results are given in Table 16. It is seen that the agreement is good, indeed surprisingly good, in view of the extremely simple assumptions made. This model deals only with the problem of a molecule escaping from a single neighbour. In section 5.2.9.2 we shall consider this behaviour in somewhat greater detail and, using a different model, establish essentially the same results.

Table 16. *Critical temperatures*

| Gas | $\Delta\varepsilon$ (J) (theoretical) | $\frac{\Delta\varepsilon}{k}$ (K) | $T_c$ (K) (observed) |
|---|---|---|---|
| He | $0.8 \times 10^{-22}$ | 6 | 5.2 |
| H$_2$ | $4 \times 10^{-22}$ | 30 | 33·1 |
| N$_2$ | $13 \times 10^{-22}$ | 100 | 121 |

### 5.1.5 *The miscibility of liquids and gases*

The continuity of the liquid and gaseous states has an interesting bearing on the solubility or miscibility of liquids. Our starting point is that, unlike solids and liquids, all gases are completely soluble in one another (see p. 57). Let us now consider the behaviour of, say, water and oil: these are mutually insoluble in the liquid state and if mixed together will separate into two distinct phases. However, water vapour and hydrocarbon vapour form a homogeneous mixture in the gaseous phase, in which there is a perfect random mixing of molecules of the two materials. If now the vapour is very highly compressed it can be brought to a density of the ordinary liquid; but provided the temperature is high enough – above the 'effective critical temperature' of the system – it will behave as a gas and still remain as a single-phase material. We shall thus have achieved complete miscibility of oil in water. Of course, this can be seen from a completely different angle. If water and oil are enclosed in a container at constant volume and heated, a temperature will be reached at which the thermal energy is sufficient to overcome the intermolecular attraction of like and unlike molecules – the material will then cease to consist of two separate phases and complete miscibility will have been achieved. Alternatively we may say that by increasing the temperature the entropy of the system is increased and the randomization which this implies may be sufficient to mix the materials on a molecular scale, i.e. to achieve miscibility. Is the single-phase material now a high-density gas or a high-temperature liquid? To some extent this is a matter of semantics.

## 5.2 Kinetic theory of an imperfect gas: van der Waals equation
### 5.2.1 *Derivation of van der Waals equation*

We shall now derive an equation for an imperfect gas, using the assumption that the main deviations from ideal behaviour arise from the finite size of the molecules and the forces between them.

We first assume that if real and ideal gases are in thermal equilibrium, i.e. at the same temperature, the mean translational energy of the

Figure 37. Potential energy between one molecule and a single neighbour as a function of separation.

individual gas molecules $(\frac{1}{2}m\overline{c^2})$ still implies a value of $\overline{c^2}$ given by $3RT/M$. This assumption, in effect, is a broader generalization of the discussion given on p. 53. (Note that the Boltzmann distribution law, $n_\varepsilon = N \exp(-\varepsilon/kT)$, applies only if $\varepsilon$ includes both kinetic and inter-molecular potential energy.)

*Size of molecule.* We first deal with the finite size of the molecules. We may say that the collisions with the wall of the container which determine the observed gas pressure are exactly the same as for zero-size molecules, i.e. an ideal gas, but that the available volume is reduced by some amount $b$ depending on the number and size of the molecules. We may, therefore, write for one mole

$$P(V - b) = RT. \tag{5.5}$$

There are several ways of estimating $b$.

($a$) Molecular collisions between the walls. If $\sigma$ is the effective diameter of a molecule (volume $v_m = \frac{1}{6}\pi\sigma^3$) the mean free path between collisions is reduced from $\lambda$ to $\lambda - \sigma$. Thus as the molecules travel between the walls of the vessel they will travel $(\lambda - \sigma)$ between collisions where, on the ideal gas model, they would have to travel a distance $\lambda$. Thus the number of collisions on the wall per second is increased in the ratio $\lambda/(\lambda - \sigma)$. This means that the pressure is increased by the factor $1/(1 - \sigma/\lambda)$. Instead of the ideal gas equation

$$PV = \tfrac{1}{3}Nm\overline{c^2} = RT,$$

we therefore have

$$PV = \frac{RT}{\left(1 - \dfrac{\sigma}{\lambda}\right)} \tag{5.6}$$

or

$$P\left(V - V\frac{\sigma}{\lambda}\right) = RT. \tag{5.7}$$

The correction term $b$ in equation (5.5) therefore has the value

$$b = V\frac{\sigma}{\lambda} = V\sigma(\pi\sigma^2 n)$$

$$= Vn\pi\sigma^3$$

$$= N_A 6v_m, \tag{5.8}$$

where we have used the simple value of $\lambda = 1/\pi\sigma^2 n$ derived in Chapter 3. On this model, therefore, $b$ is about 6 times the total volume occupied by the molecules.

(b) Excluded volume. If a molecule approaches another within a distance $\sigma$ between centres contact occurs. Thus each molecule appears to carry a sheath of radius $\sigma$, i.e. of volume $8v_m$ which excludes other molecules. If, therefore, we start with a container of volume $V$, the first molecule has a free volume $V$; the second a free volume of $V - 8v_m$; the third a free volume of $V - 2 \times 8v_m$; the fourth a free volume of $V - 3 \times 8v_m$. We then have to find an effective average value for all the $N_A$ molecules in the gas. The standard procedure is to take the geometric mean, i.e. the $N_A$th root of the product

$$V(V - 8v_m)(V - 2 \times 8v_m)(V - 3 \times 8v_m) \ldots,$$

i.e.

$N_A$th root of

$$V^{N_A}\left\{1 - \frac{8v_m}{V}[1 + 2 + 3 + \cdots + (N_A - 1)]\right\}$$

$$\approx V^{N_A}\left(1 - \frac{8v_m}{V}\frac{N_A^2}{2}\right). \tag{5.9}$$

One thus obtains as the average available volume a value very close to

$$V - V\left(\frac{8v_m}{V}\frac{N_A}{2}\right) = V - N_A 4v_m,$$

so that

$$b = 4N_A v_m. \tag{5.10}$$

(c) Excluded volume simplified. A much simpler approach (no less valid) is to say that each molecule carries an excluding volume of $8v_m$. But if we are concerned only with double collisions (ignoring triple and higher collisions) this means that on the average $8v_m$ is the excluded volume for a pair of molecules. The average is $4v_m$ so that once again we have

$$b = 4N_A v_m. \tag{5.11}$$

(d) Collisions with the wall. Jeans gives a subtle, rather more basic way of calculating $b$. He suggests that all molecules are surrounded by a sheath of radius $\sigma$ which excludes the other molecules. The excluded volume is therefore 8 times the volume of the molecules themselves. For one mole containing $N_A$ molecules this is therefore $8N_A v_m$. There is thus a zero chance of finding a molecule within the excluded volume and a chance

$$\frac{dV}{V - 8N_A v_m}, \tag{5.12}$$

of finding it in an element of volume $dV$ outside the excluded volume. He then considers the space available for molecules about to collide with the wall and argues that, since only the farther halves of the spherical sheaths are excluded when collisions are imminent, the chance of $dV$ being free in space is

$$\frac{V - 4N_A v_m}{V}.$$ (5.13)

The chance of a molecular collision with the wall is found by multiplying the two probabilities together. The result is

$$\frac{dV}{V} \times \frac{V - 4N_A v_m}{V - 8N_A v_m} \approx \frac{dV}{V} \frac{V}{V - 4N_A v_m}.$$

For a gas of negligible size the probability is $dV/V$ so that for the real gas the collision probability, and hence the pressure, are increased by the ratio $V/(V - 4N_A v_m)$. This is equivalent to replacing $V$ by $V - 4N_A v_m$ in the ideal gas equation so that

$$b = 4N_A v_m.$$

We have given these four different approaches to show that there is a simple numerical factor relating the excluded volume to the volume of the molecule but that its value depends on the model used to derive it. Experiments agree well with a value of about 4 but readers should realize that this is a 'good' value, not a 'correct' value.

### 5.2.2 Intermolecular forces – the van der Waals approach

We now turn to the effect of intermolecular forces using the approach proposed by van der Waals in 1873. Within the bulk of the gas the molecular forces, on an average, act symmetrically on one another so that the net effect is zero. Consequently within the bulk of the gas the molecules behave as though they were in a gas without attraction, so that the effective pressure is the same as for an ideal gas $P_{ideal}$. Near the walls of the vessel, however, the molecules have to escape from their neighbours before they collide with the wall. In overcoming this molecular attraction some kinetic energy is lost, the molecular velocity is reduced and the momentum imparted to the wall on impact and rebound is less than it would be for an ideal gas. Suppose the pressure defect is $\Delta P$. Then the observed pressure $P$ at the walls is less than $P_{ideal}$ by an amount $\Delta P$.

If we write the ideal gas equation, modified to allow for this, we have

$$P_{ideal} V_{ideal} = RT,$$

or

$$(P + \Delta P)(V - b) = RT.$$ (5.14)

The pressure defect is proportional to the number of molecules striking the surface per unit area per second, i.e. to the density $\rho$. It is also proportional to the number of molecules attracting the molecules and reducing their impact. This is also proportional to $\rho$. As a result $\Delta P$ is proportional to $\rho^2$. It could be written as $aP^2$ since $P$ is nearly proportional to $\rho$, but it turns out that it is better to write this as $a/V^2$; this correction term appearing to operate over a larger range than a term $aP^2$. We may thus regard the correction term $a/V^2$ as the internal pressure in the gas, i.e. the pressure that has to be exerted to pull the molecules away from one another in overcoming intermolecular forces. Our final relation is

$$\left(P+\frac{a}{V^2}\right)(V-b)=RT. \tag{5.15}$$

This celebrated relation is known as the van der Waals equation and is a very convenient way of describing the behaviour of a 'real' gas. The model on which it is derived is admittedly a very crude one but it is generally felt that the equation holds better than it ought. Its main merit is that it is simple and that it allows for molecular volume and molecular attraction with the use of only two parameters $a$ and $b$.

If $a$ and $b$ are determined for one mole of gas, the van der Waals equation of state for $n$ moles becomes

$$\left(P+n^2\frac{a}{V^2}\right)(V-nb)=nRT.$$

### 5.2.3 *Intermolecular forces – another approach*

The van der Waals approach assumes that the gas pressure has its 'ideal' value in the bulk of the gas and that the molecules suffer a loss of energy (and momentum) as they escape from the bulk and collide with the walls. However, this treatment (also given by Jeans as late as 1940) is somewhat misleading. We fix our attention on some plane in the gas normal to the $x$ direction and consider the transfer of momentum across it. If the plane is parallel to the wall of the vessel, whatever momentum is transferred across the plane will ultimately be communicated to the wall. This momentum is made up of two parts. The first part is that which is due to the kinetic energy of the molecules ($\frac{3}{2}kT$). The momentum flux per unit area in, say, the $x$ direction is then $2mu$ as for an ideal gas. The second part arises from the loss of energy and momentum as a molecule on the near side of the plane pulls away from all the molecules on the far side of the plane. This will produce a net reduction of momentum transfer across the plane proportional to (density)$^2$, i.e. to $V^{-2}$, exactly as in the van der Waals argument. Thus the pressure which is transmitted to the walls is again given by

$$P = P_{ideal} - a/V^2.$$

At first sight it would seem that the molecule which loses momentum by pulling itself away from molecules outside the plane will gain momentum because of the attraction of the molecules on its other side. This may be so but the overall transmission of momentum across the plane is determined by what happens at the plane itself. This may be seen in a different way if we consider the energy density of the gas as a whole. There will be kinetic energy ($\frac{3}{2}kT$) per molecule and also a potential energy term arising from the attraction between the molecules. This is negative (as for all systems involving attractive forces) so that the total energy is of the form

$$E_{total} = \text{kinetic energy} + \text{potential energy},$$

where the potential energy is a negative quantity. If we now consider a small isothermal compression (volume change $\Delta V$) we increase the energy density and this provides a measure of the work done $P\Delta V$ by the external pressure $P$. Clearly $P$ will be less than $P_{ideal}$ since the energy density is less.

The pressure defect ($a/V^2$) may be regarded as the cohesive energy of the gas. If the gas were so heavily compressed that the molecules were in close contact there would also be a repulsive term which would act in a sense opposite to that of the attractive forces, but the attractive forces would dominate and produce an extremely large pressure defect. Since there is nothing in this model which refers to the state of the medium it will apply to liquids as well as gases. Thus for water under standard atmospheric conditions the kinetic pressure term is of order 1312 atmospheres;[†] this large value is due to the large molecular density as compared with say water vapour at s.t.p. But the internal pressure is −1311 atmospheres. The net pressure exerted on the walls is simply 1 atmosphere.

Although this treatment emphasizes the conditions in the bulk of the gas, we can, of course, draw our imaginary plane wherever we wish. If we draw it at a distance from the wall equal to the mean free path the momentum transfer process resembles the model used in the van der Waals treatment. We shall use this idea in the following section and later, where we calculate the pressure defect in terms of intermolecular forces.

[†] See E. H. Kennard, *Kinetic Theory of Gases* (McGraw-Hill, 1938).

### 5.2.4 *Pressure defect: the Dieterici equation*

In the van der Waals treatment it is assumed that every molecule striking the wall has less energy than it would have had if there had been no intermolecular attraction. (Some molecules, in fact might never reach the wall.) If $E$ is the average kinetic energy of the equivalent ideal gas and $\Delta E$ is the energy loss, the pressure defect $\Delta P$, expressed as a ratio of the ideal pressure $P$, is given by

$$\frac{\Delta P}{P} = \frac{\Delta E}{E},$$

since we are assuming the same molecular density at the wall as in the bulk.

By contrast, Dieterici stresses that the temperature of the gas is everywhere constant, even near the walls, so that the Boltzmann distribution for the total energy applies to the molecules striking the wall as well as to those in the bulk. Since the wall-striking molecules have reduced total energy, the mean density $n$ is lower than the ideal number $n_0$ by the factor

$$n = n_0 \exp(-\Delta E/kT) \approx n_0(1 - \Delta E/kT),$$

so that

$$\frac{\Delta n}{n_0} \approx \frac{\Delta E}{kT}.$$

Since the gas pressure for a gas at a specified temperature is proportional to the density this again gives for the pressure defect

$$\frac{\Delta P}{P} = \frac{\Delta E}{kT} \approx \frac{\Delta E}{E},$$

since $kT$ (or $\frac{3}{2}kT$) is the average thermal energy of the molecules. This is not really a different model but a different way of calculating the effect. It implies that at a fixed temperature the same Boltzmann distribution applies to the wall molecules as to the bulk – but the density at the wall is less. This is analogous to the decrease in density of a gas in a gravitational field.

The energy needed for a single molecule to escape is proportional to the mean molecular density $n$, i.e. to $1/V$ so that $\Delta E$ is of the form $\alpha/V$. Hence for a mole of gas

$$P = P_{\text{ideal}} \exp\left(-\frac{\Delta E}{RT}\right) = P_{\text{ideal}} \exp\left(\frac{-\alpha}{VRT}\right).$$

Since

$$P_{ideal}V_{true} = RT,$$

we have, assuming that the excluded volume is $b$,

$$P(V-b) = RT \exp\left(\frac{-\alpha}{VRT}\right). \tag{5.16}$$

This is the Dieterici equation of state. A little manipulation shows that for small values of $\alpha$ this is identical with the van de Waals equation of state.

### 5.2.5 *Clausius' and Berthelot's equations of state*

There are two other equations of state which at various times have been proposed. R. Clausius (1822–88) attempted to allow for the effect of temperature on the cohesive forces and suggested in 1880

$$\left[P + \frac{a}{T(V+c)^2}\right](V-b) = RT. \tag{5.17}$$

This involves three arbitrary constants. D. Berthelot in 1897 found that if he used a different value for $a$ he could get about as good a fit with experiment by putting $c = 0$. He therefore used

$$\left(P + \frac{a}{TV^2}\right)(V-b) = RT. \tag{5.18}$$

These equations are mainly of historical interest. We shall not consider them further.

### 5.2.6 *Attraction of the walls*

We have spent some space in discussing intermolecular attraction as the cause of a pressure 'defect' in real gases. What of the attraction between the molecules and the walls of the container? Does the pressure exerted by a gas depend on the nature of the container? Will the gas pressure be higher for a polar container than for an inert one? The answer is simple. The wall can have no net effect. Attraction may increase the impulse during the collision but the impulse will be reduced by an exactly equal amount due to the extraction of momentum from the wall. The net effect will be zero.

This is shown schematically in figure 38. We draw our arbitrary plane (see section 5.2.3 above) at AA' at a distance from the wall equal to the mean free path $\lambda$. The velocity of a molecule in the bulk of the gas is, say, $c$. When it escapes from the attraction of its neighbours its velocity has fallen to $c'$ (position B). At B it comes under the attractive forces of the wall. Consider the interaction with a wall molecule X of mass $M$. Suppose

X is initially at rest. The molecule is accelerated towards X but also draws X out of the wall to X'. When collision occurs the velocity of the molecule (mass $m$) has been increased to $c''$ whilst the velocity of the wall molecule is, say, $v$. The gain in momentum of the molecule $m$ $(c'' - c')$ is exactly equal to the gain in momentum in the opposite direction $Mv$ of the wall molecule. Thus the wall molecule is given an increased blow because of the attractive forces, but the increased momentum is exactly nullified by the momentum withdrawn from the wall. The net momentum transfer is the same as if no attraction had existed. On rebound the molecule leaves with increased velocity so imparting a larger momentum to the wall molecule, but as the gas molecule returns to position B' the attractive forces decrease its velocity to its original value $c'$: at the same time they decrease the effective impulse on $M$, and it returns to the wall with the same momentum as though no attractive force had existed. The total momentum change is thus the same as if the wall exerted no force on the gas molecule. (This is simply a description of Newton's laws of momentum in terms of molecular interactions.) By the time the gas molecule reaches A' its velocity is back to $c$ and the bulk behaviour of the gas now resembles that of a perfect gas.

The problem in this section is often avoided in elementary texts. Usually there is a bland statement that, by applying Newton's laws of momentum, the molecular momentum is reversed after collision with the wall. This is, of course, perfectly correct but it is equivalent to assuming that the wall exists at ZZ. It ignores the detailed interaction which takes place in the region between ZZ and the wall molecules themselves.

Figure 38. Behaviour of a gas molecule as it escapes at A from the attraction of its neighbours and then approaches at B to within the range of action of the surface forces of the walls of the container.

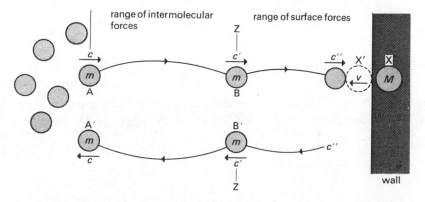

### 5.2.7 *Van der Waals equation and the Boyle temperature*

If we plot isothermals of $PV$ against $P$ we note that the Boyle temperature occurs when the slope of the curve is zero, at $P \to 0$. We therefore use the van der Waals equation, determine $\partial(PV)/\partial P$ and put it equal to zero at $P = 0$.

From the original equation (5.15) we have

$$P = \frac{RT}{V-b} - \frac{a}{V^2},$$

so that

$$PV = \frac{RTV}{V-b} - \frac{a}{V}. \tag{5.19}$$

Then

$$\frac{\partial}{\partial P}(PV) = \left[ RT\frac{1}{V-b} - RT\frac{V}{(V-b)^2} + \frac{a}{V^2} \right]\left(\frac{\partial V}{\partial P}\right)_T$$

$$= \left[ -\frac{RTb}{(V-b)^2} + \frac{a}{V^2} \right]\left(\frac{\partial V}{\partial P}\right)_T. \tag{5.20}$$

If we specify that, at $T = T_B$, this is zero when $P = 0$, this is equivalent to considering $V$ as approaching infinity so that $(V-b)$ is indistinguishable from $V$. From equation (5.20) we therefore have

$$-RT_B b + a = 0,$$

or

$$T_B = \frac{a}{Rb}. \tag{5.21}$$

At all temperatures above $T_B$, $PV$ will increase with increasing $P$, so that there will be no inflexion in the $PV$ curves. At $T_B$ itself we may substitute $T = T_B = a/Rb$ in the van der Waals equation. Multiplying out equation (5.15) we have

$$PV + \frac{a}{V} - bP - \frac{ab}{V^2} = RT.$$

On the LHS we may, if the pressures are small, replace $a/V$ by $aP/RT$, and neglect the fourth term. We are left with

$$PV + \frac{aP}{RT} - bP = RT.$$

From our value of the Boyle temperature ($T_B = a/Rb$) we see that the second and third terms on the LHS are equal and opposite. Hence at $T_B$

$$PV = RT_B$$

for a real van der Waals gas.

### 5.2.8 *Van der Waals equation and the critical points*

The van der Waals equation is a cubic in $V$. If we plot $P$ against $V$ we obtain a family of curves shown schematically in figures 39 and 40. Let us consider a typical isotherm XDBACEY in figure 39 below the critical temperature. Join D and E by an isobar which cuts the curve at A.

Consider now the behaviour at A. Suppose we started with a homogeneous phase. Along BC the slope $\partial P/\partial V$ is positive. This means that if a small increase in volume occurs this is accompanied by an *increase* in gas pressure. There is thus a spontaneous expansion towards C. Similarly a small decrease in volume would lead to a spontaneous contraction towards B. These processes constitute the initial separation into two phases, gaseous at C liquid at B. The part BD is unstable, the liquid can remain in this region (region of superheating), but generally the phase spreads to D. Similarly CE is unstable, the gas can remain in this region of supercooling but generally the phase spreads to E.† In passing we may note that some isotherms can give negative pressures, i.e. B may be below the $V$ axis. This only occurs in the liquid phase and as we shall see this provides a means of estimating the tensile strength of a liquid.

Figure 39. Stylized behaviour of a van der Waals gas.

† A thermodynamic argument quoted in J. K. Roberts, *Heat and Thermodynamics*, 5th edn (Wiley, 1960) shows the line EAD should cut the curve at such a position that the areas DBA and ACE are equal. The reasoning is as follows. In carrying the gas through a reversible cycle DBACED (i.e. from D back to D) the gas is restored to its initial state so that there is no change in entropy. Since the whole process is at a constant temperature this implies that the net work done is zero. The work done is given by the $P\,dV$ areas in each part so that the area DBA must be equal to the area ACE.

The two turning points B and C coalesce into a single region at Z on the critical isotherm (figure 40). We thus have two conditions determining the critical isotherm:

$$\left(\frac{\partial P}{\partial V}\right)_T = 0 \text{ for turning points such as B and C,}$$

$$\left(\frac{\partial^2 P}{\partial V^2}\right)_T = 0 \text{ for the coalescence of these at the point of}$$

inflexion Z.

(5.22)

Multiplying out the van der Waals equation we have

$$PV + \frac{a}{V} - Pb - \frac{ab}{V^2} = RT.$$

Differentiating with respect to $V$ at constant temperature we have

$$P + V\left(\frac{\partial P}{\partial V}\right)_T - \frac{a}{V^2} - b\left(\frac{\partial P}{\partial V}\right)_T + \frac{2ab}{V^3} = 0. \tag{5.23}$$

Inserting the condition that for all turning points $(\partial P/\partial V)_T = 0$ we obtain

$$P = \frac{a}{V^2} - \frac{2ab}{V^3}. \tag{5.24}$$

This is the equation of the curve BKZLC (figure 40). The tip of this curve corresponds to Z. This is where $(\partial P/\partial V)_T$ of equation (5.24) is zero. We have

$$\left(\frac{\partial P}{\partial V}\right)_T = -\frac{2a}{V^3} + \frac{6ab}{V^4} = 0. \tag{5.25}$$

Figure 40. The van der Waals isotherms at various temperatures.

We call the volume at which this occurs the critical volume $V_c$. From equation (5.25) we obtain

$$V_c = 3b. \tag{5.26}$$

Substituting in equation (5.24) we obtain for the critical pressure

$$P_c = \frac{a}{27b^2}. \tag{5.27}$$

Substituting $V_c$ and $P_c$ in the van der Waals equation we find the critical temperature $T_c$ from the relation

$$\left(P_c + \frac{a}{V_c^2}\right)(V_c - b) = RT_c.$$

This gives

$$T_c = \frac{8a}{27Rb}. \tag{5.28}$$

The nature of the van der Waals equation is such that all isotherms below $T_c$ have two turning-points; no isothermals above $T_c$ have even a single point of inflexion. We may note that empirically the boiling point (K) of a liquid at atmospheric pressure is about $\frac{2}{3}T_c$. There is no simple theory for this.

### 5.2.9 *Magnitudes of b and a in the van der Waals equation*
5.2.9.1 *Meaning of b.* We have already seen that for a van der Waals gas $b$ is four times the volume of the molecules, treating them as spheres of molecular diameter $\sigma$. In a mole therefore

$$b = 4N_A \tfrac{4}{3}\pi\left(\frac{\sigma}{2}\right)^3 = \frac{2\pi}{3}N_A\sigma^3. \tag{5.29}$$

From a study of the compressibility of a gas we can find the best values of $b$ to fit the data and hence calculate $\sigma$ using equation (5.29). Some typical results are given in Table 17; they are in good agreement with values of $\sigma$ deduced from diffusion, viscosity, etc.

Table 17. *Values of $\sigma$ deduced from b*

| Gas | He | $H_2$ | $N_2$ | $CO_2$ | $SO_2$ |
|---|---|---|---|---|---|
| $\sigma$ (nm) | 0.266 | 0.276 | 0.314 | 0.324 | 0.355 |

5.2.9.2 *Meaning of a.* We use here the simple van der Waals approach. The pressure of a gas on a container is due to molecular collisions. If the velocity component normal to the container wall is $u$, then as we saw in Chapter 3 we may say:

$$\text{Number of collisions } m^{-2} s^{-1} = \frac{nu}{2}$$

$$\text{Momentum change per collision } = 2mu. \tag{5.30}$$

The rate of change of momentum $m^{-2} s^{-1}$ which is the product of these is the ideal gas pressure

$$P_{\text{ideal}} = nmu^2 = \tfrac{1}{3}nmc^2. \tag{5.31}$$

In escaping from the attractive force of the last few neighbouring molecules the collision velocity is reduced from $u$ to $u - \Delta u$, so that the momentum transfer is reduced to $2m(u - \Delta u)$. The number of collisions $m^{-2} s^{-1}$ which we may call the 'flux' is still $nu/2$. At first sight it might seem that this should also be diminished to $2(u - \Delta u)/n$. This, however, is not so. Although the molecules are retarded they are not, *in this model*, prevented from reaching the wall, so continuity in flux from the bulk of the gas to the wall is maintained. The flux does, of course, depend on the molecular velocity but near the wall where retardation occurs there is a slight 'crowding-up' of molecules, i.e. a compensating increase in density of the gas. In this way flux-continuity is maintained. This model should be contrasted with the Dieterici treatment described above. The observed pressure $P$ therefore becomes

$$P = \frac{nu}{2} 2m(u - \Delta u) = nmu^2 \left(1 - \frac{\Delta u}{u}\right)$$

or

$$P = P_{\text{ideal}} \left(1 - \frac{\Delta u}{u}\right). \tag{5.32}$$

Clearly $P_{\text{ideal}} \Delta u / u$ is the van der Waals term $a/V^2$. We may write this as

$$P_{\text{ideal}} \frac{\Delta u}{u} = nmu^2 \frac{\Delta u}{u}$$

$$= \Delta(\tfrac{1}{2}mu^2)n.$$

Now $\Delta(\tfrac{1}{2}mu^2)$ is the loss in kinetic energy per molecule in the $u$ direction and $n = N/V$. Consequently the van der Waals term

$$\frac{a}{V^2} = \frac{N_A}{V} \text{ (loss of kinetic energy per molecule)} \tag{5.33}$$

or

$$a = N_A V \text{ (loss of kinetic energy per molecule)}, \qquad (5.34)$$

assuming no changes occur on the average in the $v$ or $w$ directions.

We could now assume some law of force between the molecules and integrate for all molecules and all distances, extending from the position of the last collision near the wall to all the molecules throughout the bulk of the gas. Indeed the van der Waals treatment is basically concerned with relatively long-range forces. However, as we are interested only in an order of magnitude calculation we can arrive at a reasonable value of $a$ by considering only the nearest neighbours. Because of the simple way this is done it actually overestimates the effect of the attractive forces. The method is illustrated in figure 41. Every molecule within the hemispherical shell bounded by $r_1 = \sigma/2$ and $r_2 = 3\sigma/2$ is a neighbour which actually touches the molecule considered. The work done in dragging the molecule away from these will be this number of neighbours multiplied by $\Delta\varepsilon$. The volume of the shell is $\frac{1}{2}\frac{4}{3}\pi(r_2^3 - r_1^3) = \frac{1}{6}\pi 13\sigma^3 = 13v_m$, where $v_m$ is the theoretical volume of a single molecule. The number of neighbours is then $n \times 13v_m$ so that the energy lost in pulling away the molecule, i.e. its loss of kinetic energy in the $u$ direction is

$$(\tfrac{1}{2}mu^2) = n \times 13 \times v_m \times \Delta\varepsilon.$$

The van der Waals term is then

$$a = N_A Vn\, 13v_m(\Delta\varepsilon). \qquad (5.35)$$

Since

$$b = 4N_A v_m \text{ and } N = Vn,$$

Figure 41. The range of attraction of surrounding gas molecules as a single gas molecule collides with the wall.

we have

$$a = \frac{13}{4} N_A b (\Delta \varepsilon).$$ (5.36)

Inserting this in our value for $T_c$,

$$T_c = \frac{8a}{27Rb} = \frac{8}{27} \frac{13}{4} N_A b (\Delta \varepsilon) \frac{1}{Rb} = \frac{26}{27} N_A \frac{\Delta \varepsilon}{R} = \frac{26}{27} \frac{\Delta \varepsilon}{k},$$

i.e.

$$kT_c \approx \Delta \varepsilon.$$ (5.37)

This extremely crude model gives a reasonable value for $a$ and explains why the critical temperature is nearly equal to $\Delta \varepsilon / k$. It must be emphasized that this is an 'order of magnitude' calculation and that the final constant in front of $\Delta \varepsilon$ ($\frac{26}{27} \approx 1$) is largely fortuitous. The model itself, however, is basically valid.

## 5.3 Some properties of the critical point
### 5.3.1 Critical volume
$V_c = 3b = 12 \times$ volume of molecules themselves. Hence the mean separation between the centres of the molecules at the critical point is

$$1 = (12)^{\frac{1}{3}} \sigma \approx 2.4 \sigma.$$ (5.38)

There is just about enough room between molecules to squeeze another.

### 5.3.2 Mean free path
Using our simple derivation,

$$\lambda = \frac{1}{\pi n \sigma^2},$$

$$\frac{\lambda}{\sigma} = \frac{1}{\pi n \sigma^3} = \frac{1}{6 n v_m}.$$

But $n v_m$ is the volume occupied by the molecules in $1 \, \text{m}^3$ of space. At the critical point this is $\frac{1}{12} \, \text{m}^3$. Hence

$$\frac{\lambda}{\sigma} = \frac{1}{6 \times \frac{1}{12}}$$

$$= 2.$$ (5.39)

Thus $\sigma$, $\lambda$ and the mean molecular separation are all roughly equal at the critical point. For ideal gases at ordinary temperatures and pressures the relevant quantities are roughly 0.3, 100 and 3 nm, respectively.

### 5.3.3 *Critical coefficient*

The ratio $RT_c/P_cV_c$ is known as the critical coefficient. For all equations of state which involve only two arbitrary constants the critical coefficient is an explicit dimensionless number. The values obtained for the Berthelot, Dieterici and van der Waals equations are summarized in Table 18. We see that, in general, for a simple gas $RT_c/P_cV_c$ lies between 3 and 3.5 if the attractive forces are of a van der Waals nature. For this particular parameter the Dieterici equation gives better agreement than the van der Waals equation. For larger and more polar molecules the critical coefficient is generally greater than 4.5.

## 5.4 **Law of corresponding states**

We first consider how, on dimensional grounds, we could formulate the equation of state of a real gas if we knew the law of force between the molecules as a function of separation. For an intermolecular law of force of a given shape we need two parameters to define the scale of the separation axis and the scale of the potential axis. Any two parameters will do. For convenience we may choose the molecular diameter $\sigma$ and the depth of the potential well at equilibrium $\Delta\varepsilon$ (see figure 37). The other

Table 18. *Critical points and critical coefficients*

| Equation of state | $V_c$ | $P_c$ | $T_c$ | $\dfrac{RT_c}{P_cV_c}$ |
|---|---|---|---|---|
| Berthelot see equation (5.18) | $3b$ | $\dfrac{RT_c}{2b} - \dfrac{a}{9T_cb^2}$ | $\dfrac{8a}{27Rb}$ | $\frac{8}{3} \approx 2.7$ |
| Dieterici see equation (5.16) | $2b$ | $\dfrac{a}{4e^2b^2}$ | $\dfrac{a}{4Rb}$ | $\dfrac{e^2}{2} \approx 3.7$ |
| van der Waals see equation (5.15) | $3b$ | $\dfrac{a}{27b^2}$ | $\dfrac{8a}{27Rb}$ | $\frac{8}{3} \approx 2.7$ |

Experimental values for the critical coefficient are as follows:

| Gas | He | $H_2$ | $N_2$ | $CO_2$ | Water | Acetic acid |
|---|---|---|---|---|---|---|
| $\dfrac{RT_c}{P_cV_c}$ | 3.1 | 3.0 | 3.4 | 3.5 | 4.5 | 5.0 |

parameters likely to be involved are the number density $n$, the molecular mass $m$ and the thermal energy $kT$. Dimensional analysis at once shows how we can derive the gas pressure $P$. The quantities $kT/\Delta\varepsilon$ and $V/n\sigma^3$ have zero dimensions, and $\Delta\varepsilon/\sigma^3$ has the dimensions of pressure. There is no way of including $m$. Thus

$$P = \frac{\Delta\varepsilon}{\sigma^3} f\left(\frac{kT}{\Delta\varepsilon}, \frac{V}{n\sigma^3}\right)$$

or

$$\frac{P}{P^*} = f\left(\frac{T}{T^*}, \frac{V}{V^*}\right),$$

where

$$P^* = \Delta\varepsilon/\sigma^3,$$
$$kT^* = \Delta\varepsilon,$$
$$V^* = n\sigma^3.$$

Thus the reduced equation of state is the same for all gases having the same type of force-law; the characteristic pressure $P^*$, temperature $T^*$ and volume $V^*$ are related to the quantities $\sigma$ and $\Delta\varepsilon$ which fix the shape and scale of the potential curve.† The choice of the parameters $\sigma$ and $\Delta\varepsilon$ will determine the values of $P^*$, $T^*$ and $V^*$ for any particular gas under consideration. This result is very useful because it allows measurements on one gas to be applied to others of the same type since the mass of the molecule is not involved.

More specifically we may express the behaviour of a gas in terms of their critical values. We put

$$\theta = \frac{T}{T_c}, \qquad \pi = \frac{P}{P_c}, \qquad \phi = \frac{V}{V_c}, \tag{5.40}$$

then the van der Waals equation becomes

$$\left(\pi + \frac{3}{\phi^2}\right)(3\phi - 1) = 8\theta, \tag{5.41}$$

thus

$$\pi = f(\theta, \phi).$$

Similarly the Dieterici equation becomes

$$\pi(2\phi - 1) = \theta \exp\left[2\left(1 - \frac{1}{\theta\phi}\right)\right]. \tag{5.42}$$

† A good reason for choosing these two quantities is that even if the laws of force vary somewhat from one gas to another, $\sigma$ and $\Delta\varepsilon$ are the most important quantities determining molecular interactions, whatever the precise laws of force may be.

The assumption that these reduced equations of state are equally applicable to all gases is known as the 'law of corresponding states'. As the dimensional analysis shows, this is only valid if the intermolecular law of force has the same shape for all gases being considered. It implies that, by a simple extension or contraction of the scales on which $P$ and $V$ are plotted, all gases will give identical isotherms at the appropriate corresponding temperatures.

There is one beautiful extension of the dimensional argument due to de Boer. With very light atoms or molecules one might expect quantum effects to become important. In that case $m$ comes into the picture in the form of the dimensionless parameter $h/(\sigma^2 m \Delta \varepsilon)^{\frac{1}{2}}$, where $h$ is Planck's constant. In this way he was able to predict (correctly) the critical temperature of $He^3$ before the measurements had been made.[†]

## 5.5 Internal energy and specific heat capacities of a van der Waals gas

Consider a gas the molecules of which have $f$ degrees of freedom. When the molecules are far apart we may write the internal (kinetic) energy per mole as

$$U = (f/2)RT. \tag{5.43}$$

As the molecules come closer together the attractive forces between them lower their potential energy (see 5.6.2 below) by the amount

$$\int_\infty^V (a/V^2)\, dV = -a/V. \tag{5.44}$$

The total internal energy is the sum of the kinetic and potential energies so that

$$U = (f/2)RT - a/V, \tag{5.45}$$

where the first term on the RHS is the kinetic and the second the potential energy.

We may now calculate the specific heat capacity (per mole) when the volume is kept constant:

$$C_V = \left(\frac{\partial U}{\partial T}\right)_V = (f/2)R. \tag{5.46}$$

To calculate the specific heat capacity when the pressure is kept constant, i.e. $V$ is allowed to change, we note that one part comes from the change in $U$, the other from external work done, $dV$.

$$C_P\, dT = (\partial U)_P + P\, dV.$$

† See A. B. Pippard, *Forces and Particles* (Macmillan, 1977).

From (5.45) we see that

$$(\mathrm{d}U)_P = C_V\,\mathrm{d}T + (a/V^2)\,\mathrm{d}V,$$

therefore

$$C_P\,\mathrm{d}T = C_V\,\mathrm{d}T + (P + a/V^2)\,\mathrm{d}V. \tag{5.47}$$

We may derive a relation for $\mathrm{d}V$ at constant pressure from the van der Waals equation of state:

$$(P + a/V^2)(V - b) = RT.$$

Neglecting the term $ab/V^2$ we have

$$PV - Pb + a/V = RT. \tag{5.48}$$

For a change in which $P$ is constant we have

$$P\,\mathrm{d}V - (a/V^2)\,\mathrm{d}V = R\,\mathrm{d}T, \tag{5.49}$$

$$\mathrm{d}V = R\,\mathrm{d}T/(P - a/V^2). \tag{5.50}$$

Substituting into equation (5.47) we have

$$C_P\,\mathrm{d}T = C_V\,\mathrm{d}T + \left(\frac{P + a/V^2}{P - a/V^2}\right)R\,\mathrm{d}T,$$

or

$$C_P - C_V = \left(\frac{P + a/V^2}{P - a/V^2}\right)R \approx R\left(1 + \frac{2a}{PV^2}\right) \approx R\left(1 + \frac{2a}{RTV}\right). \tag{5.51}$$

## 5.6 Expansion of gases

### 5.6.1 *Reversible adiabatic expansion of a real gas*

For an ideal gas adiabatic expansion gives a temperature drop given by

$$\frac{T_2}{T_1} = \left(\frac{V_1}{V_2}\right)^{\gamma-1}. \tag{5.52}$$

For a real gas the temperature drop is greater because of the energy expended in overcoming intermolecular attraction.

### 5.6.2 *Joule expansion into a vacuum*

For an ideal gas the internal energy is independent of the volume so that there is no net temperature change. For a real gas work is done solely against the intermolecular attractions. We may consider this as work done against the internal pressure $a/V^2$, and write it as

$$W = \int_{V_2}^{V_1} \frac{a}{V^2}\,\mathrm{d}V = a\left(\frac{1}{V_1} - \frac{1}{V_2}\right), \tag{5.53}$$

where $V_1$ is the initial and $V_2$ the final volume.

### 5.6.3 *Joule–Kelvin expansion*

This is a type of expansion suggested by Lord Kelvin and carried out by him in collaboration with Joule in 1853. The purpose was to study (under more carefully controlled conditions than Joule's earlier experiments) the heat changes occurring when a gas undergoes free expansion. The general principle is illustrated in figure 42(*a*). Gas in the left-hand part of the tube is driven at a constant pressure $P_1$ through an orifice or porous plug and withdrawn on the right-hand part at a constant, lower, pressure $P_2$. Joule and Kelvin studied the temperature changes occurring after a steady state had been reached for a system that was thermally insulated. For the purpose of our treatment, however, we shall assume that heat sources or sinks can be applied so that the temperature $T$ on both sides is maintained constant.

Let $V_1$ be the volume per mole on the LHS, $V_2$ the volume on the RHS. During the passage of one mole of gas

external work done on the gas on LHS = $P_1 V_1$,

external work done by the gas on RHS = $P_2 V_2$,

hence net work done by the gas = $P_2 V_2 - P_1 V_1$.   (5.54)

Figure 42. The Joule–Kelvin expansion. (*a*) the gas is driven at constant pressure $P_1$ through a porous plug and withdrawn at a lower pressure $P_2$. (*b*) Isotherms showing that for an expansion from a higher to a lower pressure, the change in $PV$ is positive at low temperature and negative at high temperature.

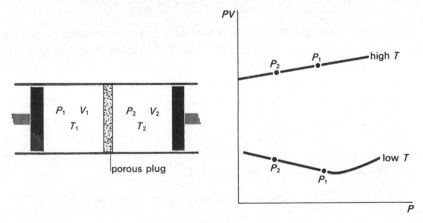

(a)           (b)

There is also work done by the gas against the internal forces. We have already calculated this in equation (5.53). The amount is

$$\frac{a}{V_1} - \frac{a}{V_2}.$$ (5.55)

Hence the total work done by the gas when one mole flows is

$$P_2 V_2 - P_1 V_1 + \frac{a}{V_1} - \frac{a}{V_2} = \left(P_2 V_2 - \frac{a}{V_2}\right) - \left(P_1 V_1 - \frac{a}{V_1}\right).$$ (5.56)

If there is to be no cooling of the gas, this amount of heat has to be supplied. We may calculate the amount of heat involved assuming the gas obeys the van der Waals equation. We start with

$$\left(P + \frac{a}{V^2}\right)(V - b) = RT,$$

and multiply out. Neglecting the term in $ab/V^2$ we have

$$PV = RT - \frac{a}{V} + bP,$$ (5.57)

or

$$PV - \frac{a}{V} = RT - \frac{2a}{V} + bP.$$ (5.58)

In the second term on the RHS we replace $V$ by $RT/P$, so that

$$PV - \frac{a}{V} = RT - P\frac{2a}{RT} + bP.$$ (5.59)

If $T$ is to be maintained constant

$$\left(PV - \frac{a}{V}\right)_2 - \left(PV - \frac{a}{V}\right)_1 = (P_1 - P_2)\left(\frac{2a}{RT} - b\right).$$ (5.60)

Since $P_1 > P_2$ the heat supplied to maintain $T$ constant must be positive if $2a/RT - b$ is positive, i.e. if $2a/RT > b$. This means that if heat is *not* supplied there will be *cooling*.

Experiments show that with nearly all gases there is a cooling at room temperature after a Joule–Kelvin expansion. With hydrogen, however, there is a rise in temperature unless the gas is first cooled to about −80 °C.

If the system is thermally isolated we may estimate the temperature difference produced during a Joule–Kelvin expansion in the following way. We write equation (5.59) more fully as

$$\left(PV - \frac{a}{V}\right)_2 - \left(PV - \frac{a}{V}\right)_1 = RT_2 - RT_1 + (P_1 - P_2)\left(\frac{2a}{RT} - b\right).$$ (5.61)

We cannot complete this rigorously without the use of the second law of thermodynamics. It turns out that the LHS, which is the net work done by the gas, is equal to $C_V \Delta T$, where $\Delta T$ is the temperature drop $T_1 - T_2$. Equation (5.61) becomes

$$C_V \Delta T = -R \Delta T + (P_1 - P_2)\left(\frac{2a}{RT} - b\right).$$

Since $C_V \approx C_P - R$ we have finally for the cooling

$$\Delta T = \frac{1}{C_P}(P_1 - P_2)\left(\frac{2a}{RT} - b\right). \tag{5.62}$$

We see that there is a critical temperature below which there is a cooling, above which a heating of the gas. This inversion temperature occurs when

$$\frac{2a}{RT_i} = b$$

or

$$T_i = \frac{2a}{Rb} = 2T_B. \tag{5.63}$$

### 5.6.4 *The reason for an inversion temperature*

It is instructive to consider separately the two main terms involved in the cooling of a gas subjected to a Joule–Kelvin expansion. If the gas expands from $P_1$ to $P_2$, i.e. if $P_1$ is greater than $P_2$, the work done by the gas which may cause cooling is

$$P_2 V_2 - P_1 V_1 + \left(\frac{a}{V_1} - \frac{a}{V_2}\right)$$

$$= \Delta[PV]_1^2 + \text{work against internal attraction.} \tag{5.64}$$

For a real gas in which there is molecular attraction the second term on the right must be positive and therefore always lead to cooling. This is not necessarily true of the first term. At low temperatures, in moving from $P_1$ to $P_2$, the quantity $\Delta[PV]_1^2$, is positive (see figure 42(b)). Consequently additional work must be done by the gas in expanding and this causes increased cooling. At high temperatures, however, well above the Boyle isotherm, $\Delta[PV]_1^2$ is negative, i.e. work is done on the gas as it expands and this opposes the cooling effect of the second term in equation (5.64). At some intermediate temperature the two effects just balance and there is neither heating nor cooling. This is the inversion temperature $T_i$.

Although this is helpful in showing how the inversion temperature arises it is, in some ways, artificial since it attempts to separate the 'imperfect' properties of the gas into two independent characteristics. As the van der Waals equation shows these two characteristics are, in fact, linked.

### 5.6.5 *An isoenthalpic expansion in the Joule–Kelvin experiment*

When a gas is expanded reversibly through nozzles or a porous plug under conditions of thermal insulation we have by the first law of thermodynamics

$$\Delta Q = 0 = U_2 - U_1 + P_2 V_2 - P_1 V_1,$$

so that

$$U_2 + P_2 V_2 = U_1 + P_1 V_1$$
$$H_2 = H_1.$$

The porous-plug expansion is thus an isoenthalpic process. The temperature change associated with the pressure change must therefore be expressed in the form $(\partial T/\partial P)_H$. For a small change in $H$ we have

$$dH = \left(\frac{\partial H}{\partial T}\right)_P dT + \left(\frac{\partial H}{\partial P}\right)_T dP. \tag{5.65}$$

If we are specifying that $H$ is constant, $dH = 0$ and we are left with

$$\left(\frac{\partial H}{\partial T}\right)_P + \left(\frac{\partial H}{\partial P}\right)_T \left(\frac{\partial P}{\partial T}\right)_H = 0.$$

Figure 43. Curves of constant enthalpy as a function of temperature and pressure (from F. W. Sears, *Thermodynamics, the Kinetic Theory of Gases and Statistical Mechanics*, 2nd edn (Addison-Wesley, 1953)).

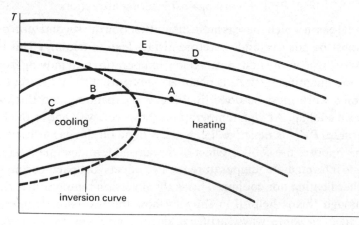

The first term on the left is $C_P$. Consequently

$$\left(\frac{\partial T}{\partial P}\right)_H = -\frac{1}{C_P}\left(\frac{\partial H}{\partial P}\right)_T. \tag{5.66}$$

In the next step we have to make use of thermodynamic functions based on the second law, viz:

$$\left(\frac{\partial H}{\partial P}\right)_T - V = -T\left(\frac{\partial V}{\partial T}\right)_P. \tag{5.67}$$

The Joule–Kelvin temperature change is therefore given by

$$\left(\frac{\partial T}{\partial P}\right)_H = -\frac{1}{C_P}\left(\frac{\partial H}{\partial P}\right)_T = \frac{T\left(\frac{\partial V}{\partial T}\right)_P - V}{C_P},$$

or

$$\Delta T = \frac{1}{C_P}\Delta P\left[T\left(\frac{\partial V}{\partial T}\right)_P - V\right]. \tag{5.68}$$

Using equation (5.57) a little manipulation shows that the term $T(\partial V/\partial T)_P - V$ in equation (5.68) is equal to $2a/RT - b$. Thus equation (5.68) becomes identical with the relation given in equation (5.62) for the Joule–Kelvin cooling. Consequently the inversion temperature is again given by

$$T_i = \frac{2a}{Rb}.$$

The above treatment suggests that there is a single inversion temperature which is independent of pressure. This results from assuming that the pressures are low, so that $ab/V^2$ can be neglected. If the higher terms are included it may be shown that for the van der Waals equation of state, $T_i$ is a parabolic function of the pressure – indeed at pressures above $a/3b^2$ the gas expansion shows inversion at all temperatures, i.e. at high pressures it will heat up on expanding.

Table 19. *Inversion temperatures $T_i$ of some gases compared with $2T_B(K)$*

| Gas | $T_B$ | $2T_B$ | $T_i$ |
|-----|-------|--------|-------|
| He | 16 | 32 | 25 |
| H$_2$ | 114 | 228 | 190 |
| N$_2$ | 420 | 840 | 650 |

This is illustrated schematically in figure 43, where curves of constant enthalpy have been plotted as a function of temperature and pressure. In expanding from say point D to point E the temperature of the gas will rise – whereas an expansion from point A or point B to point C will produce a drop in temperature.

In effect the temperature $T_i$ given in equation (5.63) for a low pressure expansion is the highest value of the inversion temperature. Some approximate values of $T_i$ at low pressures are given in Table 19 and compared with $2T_B$.

# 6

## *The solid state*

We pass at once from gases to solids. In gases the atoms and/or molecules are almost completely free; in solids they are almost completely lacking in mobility. Indeed in a solid the thermal motion is so greatly reduced that individual atoms can move from their fixed positions only with the greatest difficulty. It is this which imparts to solids their most characteristic macroscopic property – they maintain whatever shape they are given, they have appreciable stiffness. Although the atoms are in fixed locations they possess some thermal energy – in some cases individual atoms can diffuse through the solid but in general the process is extremely slow. Their main thermal exercise consists in vibrating about an equilibrium position.

### 6.1 **Types of solids**

There are three main types of solids: crystalline, glasses and polymers.

### 6.1.1 *Crystalline solids*

The main feature here is long-range order. The molecules or atoms are in regular array over extended regions within each individual grain or crystal. With large single crystals the regular array may extend over an enormous number of atoms or molecules. For example with a single crystal of copper of side 1 cm, along any one direction there will be 30 000 000 copper atoms in regular array, the arrangement of the last few being (in the ideal case) in perfect step with the first.

In spite of this extraordinary degree of regularity the individual atoms can vibrate over relatively large amplitudes about their equilibrium positions. Thermal vibrations can easily exceed one-tenth of the mean spacing; yet the mean spacing is precise and determinate to an extremely high degree of reproducibility.

Some common crystalline structures are shown in figure 44. Parts (*a*), (*b*), (*c*) are the structures most commonly observed with metals. Figure 44(*a*) is the face-centred cubic (f.c.c.) structure; this is the structure of metals such as aluminium, copper, gold, silver, lead and most of the inert gases. It is a close-packed structure in which each atom has 12 nearest neighbours. Figure 44(*b*) is the hexagonal close-packed (h.c.p.) structure and again each atom has 12 nearest neighbours. Amongst the more common metals it is represented by cadmium and zinc. Figure 44(*c*) is the

Figure 44. (*a*), (*b*) and (*c*): the common structures in metals. (*a*) f.c.c. with 12 nearest neighbours; (*b*) h.c.p. with 12 nearest neighbours; (*c*) b.c.c. with 8 nearest neighbours. (*d*), (*e*) and (*f*): typical structures in ionic crystals of the type $X^+Y^-$ or $X^{2+}Y^{2-}$. (*d*) CsCl structure in which each ion of one type is surrounded by 8 ions of the opposite sign; (*e*) NaCl structure in which each ion is surrounded by 6 of the opposite sign; (*f*) ZnS (zinc blend) structure in which each ion is surrounded by 4 ions of the opposite sign at the corners of a tetrahedron. If all the atoms in (*f*) are of the same type we have the diamond structure, for example silicon, germanium, diamond.

(a)

(b)

(c)

• Cs
o Cl

(d)

• Na
o Cl

(e)

• Zn
O S

(f)

body-centred cubic (b.c.c.) structure. This is less closely packed than f.c.c. and h.c.p., each atom having only 8 nearest neighbours. The commonest examples of these are the alkali metals, $\alpha$-iron, and a number of the transition metals such as molybdenum and niobium. Figures 44($d$), ($e$) and ($f$) show the three main crystalline arrangements of equi-valent ionic crystals of the type $X^+Y^-$ or $X^{2+}Y^{2-}$. The structures depend on the ratios of the effective radii of the positive and negative ions and represent a subtle balance between the energy of interaction of the ions and their packing. Figure 44($d$) is the CsCl structure which occurs when $r_+/r_-$ is greater than 0.732 and less than 1. The coordination number is 8, i.e. each ion is surrounded by eight ions of opposite sign. Figure 44($e$) is the NaCl structure which occurs when $r_+/r_-$ is greater than 0.414 and less than 0.732: the coordination number is 6. Finally figure 44($f$) is the ZnS (zinc blend) structure which occurs when $r_+/r_-$ is greater than 0.225 and less than 0.414. The coordination number is 4. These details are summarized in Table 20. We may also note that if the two atoms in the zinc-blend structure are identical we have the diamond structure. This is represented by silicon, germanium and (naturally) diamond. Each atom has four nearest neighbours in tetrahedral array. The structure may also be regarded as being equivalent to two interpenetrating f.c.c. lattices.

### 6.1.2 *Main types of bonding in crystalline solids*
We summarize four main types of bonding and the corresponding crystalline states in Table 21.

### 6.1.3 *Glasses*
Here the individual atoms or molecules are in ordered array over a short range but there is no long-range order. They are indeed, in structure, very much like an instantaneous snap-shot of a liquid and are

Table 20. *Structure, coordination number and Madelung constant of ionic crystals*

| Ionic ratio $r_+/r_-$ | Structure | Coordination number | Madelung constant |
|---|---|---|---|
| $0.732 < r_+/r_- < 1$ | CsCl | 8 | 1.763 |
| $0.414 < r_+/r_- < 0.732$ | NaCl | 6 | 1.748 |
| $0.225 < r_+/r_- < 0.414$ | ZnS | 4 | 1.638 |

sometimes called supercooled liquids. However, molecular movement, a characteristic feature of liquids, is minute and the attempt to describe their flow properties in terms of an equivalent viscosity leads to astronomical values of the viscosity. For example from the viscosity of certain silicate glasses at elevated temperatures it is possible to extrapolate and deduce a room temperature value of the order of $10^{69}$ Pa s.

An American physicist has pointed out that this is equivalent to stating that if the glass were in the form of a rod of length equal to the circumference of the Universe and was subjected to half its breaking tension for a period of time equal to the age of the Universe, its total extension could amount to less than one thousandth of the diameter of an electron.[†] For such materials, room temperature viscosity is clearly not a very useful physical concept.

Glasses may also be described as amorphous solids.

Table 21. *Main types of bonding*

| Bonding | Bond strength | Examples | Crystalline state |
|---------|---------------|----------|-------------------|
| Van der Waals | Weak | Solid $H_2$, Kr paraffins | Close packing of weakly attracting units |
| Ionic | Strong | NaCl crystal | Giant aggregates of positive and negative ions closely packed in a way consistent with neutrality of charge |
| Covalent | Strong | Diamond, Si, Ge | Giant molecules with directed bonds, packing determined by valency number and valency directions |
| Metallic | Strong | Metals | Metal atoms give up their valency electrons leaving metal ions in a sea of electrons. Forces between ions and electrons are central giving close packing. Strong attraction gives strength. Mobile electrons give conductivity. |

[†] F. W. Preston, *Journal of Applied Physics*, vol. 13 (1942), p. 623.

### 6.1.4 *Rubber or polymers*

These consist of long organic molecules held together either by chemical cross links, by hydrogen bonds or simply by van der Waals forces. Below a critical temperature (the glass transition temperature) they behave very much like glasses, but with a greater degree of ductility; above this temperature they are generally rubber-like (see Chapter 7, section 7.4).

## 6.2 **Solid–liquid transitions**

If we heat a solid, the molecules or atoms acquire sufficient energy to escape from their lattice sites and ultimately melt, the temperature remaining constant until the whole of the specimen is molten. The quantity of heat required to convert a unit mass of the solid at its melting point to a liquid is called the latent heat of fusion.

The melting point of a solid depends on the pressure to which it is subjected. For a solid which is more dense than the liquid (this is the usual case) the melting point increases with pressure: for a solid which is less dense (e.g. ice) the melting point decreases with pressure (see figure 45(a) and (b)). If we combine this with the liquid–vapour transition we obtain a

Figure 45. Melting curves (solid–liquid equilibrium) as a function of pressure for (a) a solid more dense than the liquid, (b) a solid less dense than the liquid.

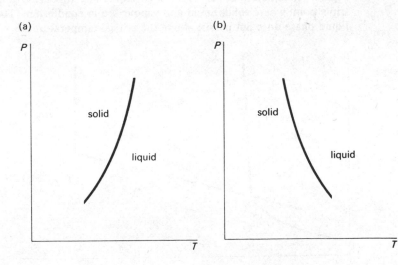

phase diagram as shown in figure 46. We note the following characteristics:

(*a*) for the liquid–vapour phase equilibrium the curve ends at C.
Above this point, which corresponds to the critical temperature $T_c$, the
liquid ceases to exist;

(*b*) by contrast, for the solid–liquid phases there is no evidence that
there is an end-point to OY above which only one phase can exist.

## 6.3 Consequences of interatomic forces in solids

We describe here a few of the properties of solids which are
directly explicable in terms of interatomic forces.

### 6.3.1 *Heat of sublimation*

If we could take a van der Waals solid such as solid argon at
absolute zero, and heat it until all the constituent atoms evaporate we
should have pumped into the solid sufficient energy to break the bonds
between all the atoms. This is the theoretical heat of sublimation $L_S$. It is
equal to the latent heat of fusion ($L_F$) plus the latent heat of vaporization
($L_V$) plus a specific heat capacity term involved in raising the temperature
from absolute zero to the temperature of vaporization. If there were an
exact energy balance the released atoms would just leave the solid with a
minute escape velocity; in practice they will have considerable thermal
energy. Thus precise comparisons between bond strengths and energy of

Figure 46. Equilibrium curves for solid, liquid and vapour. O is the
triple point where solid, liquid and vapour are in equilibrium. The
liquid phase does not persist above the critical temperature.

sublimation is difficult. However, as an order of magnitude calculation such correlation is easy, as we shall now show.

If the potential energy between each atom (or each molecule) and its neighbour is $\Delta\varepsilon$, and if each atom (or molecule) has $n$ nearest neighbours, the energy required to break the bonds in one mole will be

$$L_S = \tfrac{1}{2}N_A n \Delta\varepsilon. \tag{6.1}$$

This is allowing only for nearest-neighbour interactions, where $N_A$ is Avogadro's number and the factor $\tfrac{1}{2}$ is introduced to avoid counting each bond twice. For close packing, such as in an f.c.c. structure or an h.c.p. structure, $n$ is 12.

Hence

$$L_S = 6\Delta\varepsilon N_A. \tag{6.2}$$

If $\Delta\varepsilon$ is in Joules and we wish to express $L_S$ in kJ per mole we obtain

$$L_S \approx 36 \times 10^{20}\, \Delta\varepsilon. \tag{6.3}$$

Some typical results are given in Table 22, the 'observed' values of $L_S$ being simply the sum of $L_F$ and $L_V$. The values of $\Delta\varepsilon$ are obtained either from theoretical calculations of the van der Waals forces, or from the behaviour of the substance in the gaseous state (e.g. the deviation from ideal gas behaviour at high pressures).

Apart from helium, where the effects of 'zero-point energy' are important, the agreement is very satisfactory. A more detailed consideration of lattice energy, latent heat of sublimation and intermolecular forces will be given in the next chapter. At this stage we merely draw attention to the direct agreement with the simplest types of solids. We may also note that the latent heat of fusion is generally much lower than

Table 22. *Heats of sublimation*

| Substance | $\Delta\varepsilon$ (J) | $L_S$ (kJ mol$^{-1}$) | |
| --- | --- | --- | --- |
| | | Calculated | Observed |
| He | $0.9 \times 10^{-22}$ | 0.33 | 0.08 |
| Ne | $4.8 \times 10^{-22}$ | 1.7 | 1.3 |
| Ar | $16.5 \times 10^{-22}$ | 5.9 | 7.3 |
| Kr | $25 \times 10^{-22}$ | 9.0 | 10.0 |
| H$_2$* | $4 \times 10^{-22}$ | 1.4 | 1.0 |
| N$_2$* | $13 \times 10^{-22}$ | 4.7 | 5.4 |

* Not f.c.c. structure.

the latent heat of vaporization. From the point of view of *thermal energy* the liquid state, near the melting point, is much nearer to the solid state than it is to the gaseous state.

### 6.3.2 *Surface energy*

The atoms in the bulk of a solid are subjected to the attraction of atoms all round them, but at the free surface they are subjected only to the inward pull of the atoms within the solid. Consequently the surface layers have a higher energy than those in the bulk and this excess is called the free surface energy $\gamma$.

We may calculate this in terms of the bond energy for, say, a typical f.c.c. solid where all atoms except those in the surface have 12 nearest neighbours. An atom X in the surface has only 8 nearest neighbours. This is shown in figure 47($a$). If the crystal contains a total of $N$ atoms and the surface contains $N_s$ atoms the total bond energy of the crystal, counting only nearest-neighbour interactions, is then

$$\tfrac{1}{2}\Delta\varepsilon[12(N-N_s)+8N_s]=\tfrac{1}{2}\Delta\varepsilon[12N-4N_s]. \tag{6.4}$$

(This assumes that there is no change in lattice spacing in the surface layers; in fact a small change in spacing does occur.)

If there were no surface atoms the energy would be $\tfrac{1}{2}\Delta\varepsilon\times12N$. Thus the energy is reduced by $\tfrac{1}{2}\Delta\varepsilon\times4N_s$. Since $\Delta\varepsilon$ is negative this is a positive increase in energy which is indeed the surface energy. If the surface area of the crystal is $A$,

$$\gamma A=\tfrac{1}{2}\Delta\varepsilon\times4N_s. \tag{6.5}$$

Figure 47. f.c.c. crystal of side $2a$. ($a$) The surface atom X has 8 nearest neighbours at B, B, B, B and C, C, C, C; whereas in the bulk each atom has 12 nearest neighbours. ($b$) If the crystal is pulled apart in the plane PQRS each atom such as X breaks 4 bonds.

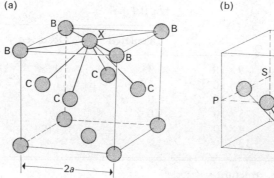

(a)                                          (b)

We may obtain this result more simply by considering a rectangular bar of crystal of cross-sectional area $A$. Any section across the crystal contains $N_s$ atoms, so if we pull the crystal in two each surface atom breaks four bonds and the work done per atom is $4\Delta\varepsilon$ (see figure 47($b$)). Hence the total work done is $4\Delta\varepsilon \times N_s$. But the area exposed is $2A$. Hence

$$2\gamma A = 4\Delta\varepsilon \times N_s. \tag{6.5a}$$

This is identical with equation (6.5). If in the crystal orientation we have considered there are $z$ atoms per m$^2$ we see from equations (6.5) and (6.5$a$) that

$$\gamma = 2 \times \Delta\varepsilon \times z. \tag{6.6}$$

But the energy necessary to sublime $z$ atoms from the bulk of the crystal (see equation (6.2)) is

$$L_S = 6\Delta\varepsilon z. \tag{6.7}$$

Hence the surface energy is equal to one-third of the heat of sublimation of a single atomic layer of atoms. For different orientations the fraction will be a little different but generally it will be of the order of $\frac{1}{3}$ to $\frac{1}{2}$. For an f.c.c. lattice of side $2a$ as shown, the volume occupied by each atom is $2a^3$. If $M$ is the molecular weight and $\rho$ the density

$$\frac{M}{N_A\rho} = 2a^3 \quad \text{or} \quad a^3 = \frac{M}{2N_A\rho}.$$

For the (100) face that we have considered the area occupied by each atom is $2a^2$. The number of atoms $z$ per m$^2$ is $1/2a^2$. Hence

$$z = \frac{1}{2a^2} = \frac{1}{2}\left(\frac{2N_A\rho}{M}\right)^{\frac{2}{3}} \tag{6.8}$$

$$= 0.8\left(\frac{N_A\rho}{M}\right)^{\frac{2}{3}}.$$

Hence

$$\gamma = \frac{0.8}{3}\left(\frac{L_S}{N_A}\right)\left(\frac{N_A\rho}{M}\right)^{\frac{2}{3}}. \tag{6.9}$$

This of course refers to a specific face and to a specific structure. For other faces and structures we should merely change the numerical factor by a small amount. A reasonable average value for the factor in front of equation (6.9) is 0.3. This relation should apply best to van der Waals solids such as solid neon, argon and krypton. However, the surface energy of such materials is not known. We have therefore applied the model to metals, the results being given in Table 23. The quantitative agreement between the calculated and observed values is poor, but the agreement in relative values is surprisingly good. If all the calculated values are multiplied by a correction factor of 0.4 the calculated values all agree with observation to within about 10 per cent. The basis for this correction factor is discussed in the next chapter.

With liquids, the surface energy is relatively easy to determine; with solids, far more difficult. In the zero-creep method for metals a series of wires of fixed radius $r$ is allowed to deform at high temperatures under different tensile loads $W$. Heavily loaded wires extend: lightly loaded wires contract due to the surface energy. For a notional extension $dl$ the virtual work $Wdl = \gamma \, dA + A d\gamma$ where $A = 2\pi rl$. Zero creep occurs at load $W_0$ and here $d\gamma = 0$. Assuming constant volume, $lr^2 = $ constant therefore $l2rdr + r^2 \, dl = 0$ so that $dA = \pi rdl$. Hence $W_0 = \pi r\gamma$ or $\gamma = W_0/\pi r$. For most metals $\gamma$ is of order 500 to 2000 mJ m$^{-2}$.

With a 'perfectly brittle' solid like mica the surface energy can be determined by finding the force required to open a crack. If we consider a cleaved specimen as shown in figure 48($b$) we may allow the force $F$ to move and to open up the crack as shown. The work done $= \int_{y_1}^{y_2} F \, dy$; this is expended in changing the elastic strain energy in the two mica leaves and also in creating two new surfaces each of area $A$. The work done can

Table 23. *Surface energies of some metals in solid state*

| Metal | MP (°C) | $L_S$ (kJ mol$^{-1}$) at 25 °C | $M$ (kg) | $\rho$ (kg m$^{-3}$) | $\gamma$ (m J m$^{-2}$) Calc. | Obs.* |
|---|---|---|---|---|---|---|
| Indium | 156 | 240 | 0.115 | $7.3 \times 10^3$ | 1300 | 630 |
| Lead | 327 | 200 | 0.207 | $11.3 \times 10^3$ | 1200 | 560 |
| Gold | 1063 | 380 | 0.197 | $19.3 \times 10^3$ | 3300 | 1400 |
| Iron | 1537 | 400 | 0.056 | $7.9 \times 10^3$ | 4000 | 2100 |
| Platinum | 1769 | 570 | 0.195 | $21.5 \times 10^3$ | 5000 | 2500 |
| Tungsten | 3380 | 850 | 0.184 | $19.3 \times 10^3$ | 6700 | 2900 |

* Observed measurements are generally made at elevated temperatures just below the MP. Values of $\gamma$ increase as the temperature is reduced reaching a maximum at 0 K where $\gamma$ is 5 to 10 per cent greater than the values quoted in the table. See W. R. Tyson, 'Surface energies of solid metals', *Canadian Metallurgical Quarterly*, vol. 14 (1975), pp. 307–14.
*Note*: mJ m$^{-2}$ = erg cm$^{-2}$.

be measured, the elastic energy calculated and the difference may then be put equal to $2\gamma A$. For mica in a good vacuum $\gamma$ has a value of about $5000 \text{ mJ m}^{-2}$; in air about $300 \text{ mJ m}^{-2}$. In general such experiments cannot be carried out satisfactorily with ductile solids such as metals or polymers, since the plastic work absorbed by deformation at the tip of the crack may be several hundred times greater than the surface energy. For example values of $\gamma$ often come out to be of the order of $10^5 \text{ mJ m}^{-2}$. Therefore the experiments can be carried out only on brittle solids. Reasonable values have been obtained on glass (A. A. Griffith 1920) and as mentioned above on mica (J. W. Obreimov 1931, A. I. Bailey 1955). Recently Gilman has shown that the technique can be applied success-fully to many crystalline solids if the experiments are carried out at liquid air temperature – he showed that at this temperature the ductility is 'frozen out'. Surface energies of the order of a few hundred $\text{mJ m}^{-2}$ are obtained for materials such as NaCl, LiF.

We may mention here a simple feature involved in the surface energy of solids which differentiates it markedly from that of liquids. Suppose we have a surface of area $A$ possessing surface energy $\gamma$. If we increase the

Figure 48. Determination of surface energy of solids (*a*) by finding the load $W_0$ at which zero creep occurs: under these conditions $W_0 = \pi r \gamma$ (*b*) by bifurcation of a mica strip. The work done in opening up the strip is equal to the change in elastic strain energy plus the increase in surface energy ($2\gamma A$) involved in creating two surfaces of area $A$.

(a)        (b)

area by a small increment the work done per unit increase may be written

$$\frac{d(A\gamma)}{dA} = \gamma + A\frac{\partial\gamma}{\partial A}.$$  (6.10)

For liquids the term $\partial\gamma/\partial A$ is zero, since, on account of atomic mobility, the structure of a liquid surface is unchanged when the surface area is increased. This is not so with solids; when the surface is stretched (even only notionally) the atoms are pulled apart and $\gamma$ diminishes so that $\partial\gamma/\partial A$ is negative. This implies that with solids the surface energy is not the same as the surface force (more strictly the line tension in the surface; see Chapter 10 on the surface tension of liquids); with liquids they are identical.

Although one cannot equate surface energy with surface force there is generally some tension in the free surface. Even if it is as large as an effective line-tension of $1\ \mathrm{N\ m}^{-1}$, this is minute compared with the elastic modulus of the solid, usually about $10^{11}\ \mathrm{N\ m}^{-2}$. Clearly forces of this magnitude can have very little visual effect on metals; so they do not influence the shape of a solid in the same way as they do liquids. A centimetre cube of copper is not dragged into spherical shape at room temperature because of surface tension. However, with very minute particles some such effects may be observed.

Surface energy of solids plays a very important part in sintering; and it is a crucial factor in the wetting of solids by liquids. From the point of view of atomic mechanisms it is worth emphasizing that the same forces which produce the pressure defect in a real gas, are responsible for the surface energy of a solid and the surface tension of a liquid.

### 6.3.3 *Elastic stiffness*

If atoms or molecules in a solid are subjected to a stress and thereby displaced from their equilibrium positions the interatomic forces will tend to restore them to their original positions. If the distortions are not too large, when the stress is removed the atoms will (under ideal conditions) return to exactly the same situation as before; the elastic modulus is defined as the ratio of the applied stress to the deformation produced. This will be discussed in detail in a later chapter. Here we merely quote the result for Young's modulus which is defined as

$$E = \frac{\text{tensile stress increment } \Delta\sigma}{(\text{increase in length } \Delta x)/(\text{original length } x)}$$

$$= x\frac{\Delta\sigma}{\Delta x}.$$  (6.11)

The elastic modulus of a solid may be easily demonstrated if one considers the force–displacement curve. Figure 49 shows the force $F$ between two neighbouring atoms as a function of separation. At low temperatures, where thermal energy can be neglected, the equilibrium separation occurs where the attractive force equals the repulsive force. Thus OA is the value of the equilibrium separation $x$. The slope of the line ZAY through A is $dF/dx$. Hence the quantity $x(dF/dx)$ is the exact one-dimensional analogue of Young's modulus as defined above.

If the solid is heated, thermal expansion (see below) shifts the equilibrium positions from A to A'. The slope $dF/dx$ is seen to be smaller and this decrease is appreciably greater than the corresponding increase in the new equilibrium separation O'A'. Thus the effective modulus is reduced. If we tried to construct a simple cubic structure out of these pairs of atoms the number in any square metre of an atomic plane would be $1/x^2$ so that the stress (as distinct from the force) is $F/x^2$ and the strain is $dx/x$. Then Young's modulus becomes $(1/x)\,dF/dx$. In this case we see that the slight increase in $x$, with rise in temperature, augments the decrease in $dF/dx$ so that both terms of the modulus are reduced by a rise in temperature. However, as already mentioned, the change in $x$ is in general small compared with the change in $dF/dx$ since $F$ is usually a 'strong' power function of $x$.

We conclude that if the behaviour of the solid as a whole resembles that of the diatomic model discussed above the elastic modulus decreases as

Figure 49. (*a*) Force–separation diagram illustrating the decrease in elastic modulus with expansion. (*b*) Potential energy–separation diagram illustrating the increase in separation which occurs when the temperature is increased (thermal expansion).

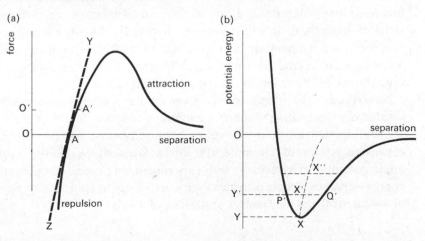

the temperature is increased. This is true of practically all solids except rubber-like solids.

### 6.3.4 *Thermal expansion*

If the length of a solid specimen at 0 °C is $l_0$ and at $t$ °C is $l_t$ one can usually write

$$l_t = l_0(1 + \alpha t),$$

where $\alpha$ is the coefficient of linear expansion and for most solids is of order $10^{-5}$ °C$^{-1}$.

This behaviour may again be explained in terms of the forces between the individual atoms. It is most convenient to plot the potential energy curve as a function of separation. If we do this for two atoms we obtain the typical curve shown in figure 49(*b*). At absolute zero the equilibrium position (ignoring zero-point energy) is at X, so the equilibrium separation is YX. As the temperature is raised the energy rises to, say, Y′ and one of the atoms will oscillate relative to the other between positions P′ and Q′ about a mean position X′. Because of asymmetry in the shape of the potential energy curve the length Y′X′ is greater than YX. Thus the mean separation increases with temperature along the curve XX′X″ . . . .

If an assembly of atoms in a solid behaved similarly this would explain thermal expansion. Of course there are added complications. In a solid one ought to treat this problem in terms of the free energy. This involves concepts beyond the scope of this book; but we shall discuss briefly the main ideas in Chapter 9, pp. 218–19. Again in the diatomic model the atoms dissociate when the energy reaches 0. This is equivalent to sublimation of the solid and tells us nothing of the melting process that occurs somewhere *en route*. This is because melting is a cooperative process involving many atoms and a satisfactory model must be multi-dimensional to allow for their behaviour (see Chapter 10). Finally we may note that increased temperature may give rise to transverse oscillations. In some types of crystals this may actually lead to a contraction in one crystallographic direction as the temperature is raised.

Nevertheless the simple model given above provides a reasonably satisfactory explanation of thermal expansion in solids. In a later chapter we shall indeed use it to derive an explicit relation for the thermal expansion in terms of intermolecular forces. We shall see that the results are in quantitative agreement with experiment. In general the potential energy–separation curve provides a very useful way of approaching many of the physical and mechanical properties of solids.

# 7

## The elastic properties of solids

If we subject a solid rod to tension we pull the atoms further apart; if we compress it we push them closer together. The interatomic forces resist these changes and, if the distortions are not too large, the body will return to its original dimensions when the external forces are removed; the deformations are then called elastic.

### 7.1 Some basic elastic properties
#### 7.1.1 Elastic moduli

If a solid rod of uniform cross-sectional area $A$ is pulled with a tensile force $F$ the tensile stress is defined as

$$\sigma = \frac{F}{A}.$$

If, as a result, the rod increases its length from $l$ to $l + \Delta l$ the fractional increase $\Delta l / l$ is called the linear strain $\varepsilon_l$ (see figure 50(a)). Within the

Figure 50. (a) Deformation in tension, (b) deformation in shear, (c) deformation under hydrostatic pressure.

elastic range Hooke showed that the stress is proportional to the strain. The ratio is defined as Young's modulus $E$:

$$E = \frac{\text{tensile stress}}{\text{linear strain}} = \frac{F/A}{\Delta l/l}. \tag{7.1}$$

A solid may also be deformed in shear. For example if we consider a rectangular element as in figure 50($b$), we may apply a shear stress $\tau$ (force ÷ area) across the faces AB and CD. This must be accompanied by equal shear stresses $\tau$ along CB and AD so that the net couple is zero (see Chapter 8, section 8.1.2). The element is then distorted through a small angle $\theta$ which is termed the shear strain. The strain is again found to be proportional to the stress and the ratio is termed the rigidity modulus $n$:

$$n = \frac{\tau}{\theta}. \tag{7.2}$$

A solid may also be deformed by a hydrostatic pressure $P$. If the solid has volume $V$, and the change in volume is $-\Delta V$, then the volume strain is $-\Delta V/V$ and this is found to be proportional to $P$. The ratio is the bulk modulus $K$.

$$K = \frac{P}{-\Delta V/V}. \tag{7.3}$$

The reciprocal of this modulus is known as the compressibility, $\beta$.

Finally, when a body is extended in tension to give a linear strain $\varepsilon_1$, it contracts in directions at right angles to the direction of tension. The fractional contraction $\varepsilon_t$ in the transverse direction is found to be proportional to $\varepsilon_1$. The ratio

$$\frac{\text{contractile strain}}{\text{linear strain}} = \frac{\varepsilon_t}{\varepsilon_1} = \text{Poisson's ratio } \nu. \tag{7.4}$$

For most solids $\nu$ has a value between $\frac{1}{4}$ and $\frac{1}{3}$ and this implies a decrease in volume under tension. If $\nu = \frac{1}{2}$ (as is nearly the case with rubber) there is no volume change in tension or compression.

There are several relationships between $E$, $n$, $K$ and $\nu$ which we shall not prove here. For example

$$n = \frac{E}{2(1+\nu)}, \tag{7.5}$$

$$K = \frac{E}{3(1-2\nu)}. \tag{7.6}$$

We see from equation (7.6) that for values of $\nu = 0.25$ to $0.33$, $K$ is comparable with $E$; for a value of $\nu = \frac{1}{2}$, $K = \infty$. This implies that the material is incompressible, which is what would be expected since tension and compression produce no volume change in such a material.

### 7.1.2 *Torsional rigidity*

If a circular tube of internal radius $r_1$, external radius $r_2$, length $l$ is held at one end and subjected to couple $\Gamma$ at the other end, every part of the tube is subjected to shear. Suppose the free end is twisted through an angle $\phi$. The ratio $\Gamma/\phi$ is known as the torsional rigidity. We shall not derive it but simply quote the relationship for $\Gamma/\phi$ to be:

$$\frac{\Gamma}{\phi} = \frac{n\pi}{2l}(r_2^4 - r_1^4), \tag{7.7}$$

where $n$ is the rigidity modulus.

### 7.1.3 *The bending of a cantilever*

If a bar is encastered at one end and deflected sideways by a force $F$ at the other end, the deflection produced depends on $F$, on the section of the bar, its length and its Young's modulus. Although we shall not derive the equation here, we shall indicate the principle of the analysis involved. Suppose figure $51(a)$ represents, in exaggerated form, the bending of a beam. The sections p, q, r, s, t indicate how the beam is deformed. We make a notional cut through the bar at section r; then the material to the left of r is subjected to a bending moment $Fx$. If the beam is to be in equilibrium this must be counterbalanced by a reverse moment of the same magnitude. The material above AB is stretched and so exerts a tensile force to the left. The material below AB is compressed and so exerts a resisting force to the right. The beam bends to just that point where these two forces produce a couple exactly equal to $Fx$. This implies that between the stretched and compressed portions there is a line AB which is unchanged in length. AB is, in fact, a part of the line XX' which represents the original length of the unstressed bar; it is known as the neutral axis. We may note in passing that the section between s and t has to carry the smallest bending moment and is therefore distorted least: the end part of the beam is indeed straight. On the other hand, the bending moment is a maximum at CD; here the curvature is a maximum. This raises an interesting point which is usually ignored by physicists and engineers. The encastered portion CDEF is considered completely rigid and subjected to zero bending moment. This implies that at the boundary CD the bending moment suddenly drops from a maximum value on the right to zero on the left. This is impossible, for there must be a transitional region HK within the encastered zone (figure 51($b$)) where the bending moment gradually falls to zero. In practice this implies a certain amount of slip between the bar and the clamp at ED and FC or some other type of yielding in the clamped zone. This is not of general importance in calculating the deflection of a loaded bar. It can, however, be of very great

Figure 51. ($a$) Elastic bending of a beam encastered at one end and subjected to a force $F$ at the other. The line XX is the neutral axis where there is neither extension nor compression of the original length. ($b$) The bending moment has a minimum at the free end and a maximum at the point where the beam enters the encastered zone. Simple theory assumes that it then drops at once to zero; in practice there must be some transition region HK. ($c$) The bifurcation region of a mica specimen resembles the encastered zone of a bent beam.

importance; for example in the propagation of a crack or in the splitting of mica described earlier as a means of determining the surface energy. The conditions at the region of bifurcation resemble those of the encastered region in a bent beam (figure $51(c)$). The analysis of the forces in this region is very difficult indeed.

If the second moment of area of the section of the beam about the vertical axis is $I$ and the length of the beam is $l$, the deflection $y$ produced at the end by a force $F$ is given by

$$y = \frac{F}{EI} \frac{l^3}{3}.$$ (7.8)

For a beam of rectangular section, thickness $a$, width $b$, this becomes

$$y = \frac{F}{E} \times \frac{4l^3}{ab^3}.$$ (7.9)

### 7.1.4 *Thermal fluctuations in the length of an elastic bar*

Consider a bar of length $L$, cross-section $A$, Young's modulus $E$. If we apply a force $F$, the stress is $\sigma = F/A$ and the strain $\varepsilon$ is then $(1/E) \times$ stress. Hence

$$F = EA\varepsilon.$$

If an incremental force $dF$ produces an incremental strain $d\varepsilon$, the incremental extension is $L\,d\varepsilon$ so that the force does work

$$FL\,d\varepsilon = EAL\varepsilon\,d\varepsilon.$$

In stretching the bar from its initial state to a final extension $\varepsilon_0$ the work $W$ done is the integral of equation

$$W = \tfrac{1}{2}EAL\varepsilon_0^2.$$

This is the elastic strain energy and will be positive whether the bar is extended ($\varepsilon +$ve) or compressed ($\varepsilon -$ve).

If as a result of thermal fluctuations the bar undergoes extension or contraction equivalent to a strain $\varepsilon_0$ the bar will possess potential energy $\tfrac{1}{2}EAL\varepsilon_0^2$ over and above its average thermal energy. This is proportional to the square of a parameter ($\varepsilon_0$) and may therefore be treated as a single degree of freedom. The average value of $\overline{\varepsilon_0^2}$ is thus given by

$$\tfrac{1}{2}EAL\overline{\varepsilon_0^2} = \tfrac{1}{2}kT$$

$$\overline{\varepsilon_0^2} = \frac{kT}{EAL}.$$

For a bar where $A = 100 \text{ mm}^2$, $L = 1 \text{ m}$, $E = 10^{11} \text{ N m}^{-2}$ (say brass) we find that at $T = 300 \text{ K}$, $\overline{\varepsilon_0^2} = 4 \times 10^{-28} \text{ m}^2$. The r.m.s. extension or contraction due to thermal fluctuations is thus of order $2 \times 10^{-14}$ m or less than a ten-thousandth of the diameter of an atom.

### 7.1.5 *Young's modulus in terms of interatomic force constant*

In the previous chapter we showed how the elastic modulus is related to the force–displacement curve. We shall now make this more quantitative. Consider a solid composed of a single species of atom and let us assume a simple cubic structure, the separation between each atom being $a$. Let the bar have unit cross-sectional area and consider three neighbouring planes I, II and III (figure 52). We assume that there are only nearest-neighbour interactions; then if an atom q in plane II is displaced in a direction normal to the plane, it experiences a restoring force only from atom p in plane I and atom r in plane III. If a tensile force $f$ is applied to atom r the spacing between the atoms in that row will each be increased by a small distance $x$ (where $x \ll a$). The restoring force which each atom in this row experiences will be proportional to $x$. Let the restoring force be $kx$, where $k$ is, in fact, a measure of the slope of the force–displacement curve shown in the previous chapter (figure 49(a)).

Since there are $1/a^2$ atoms per m$^2$, the force per m$^2$ (the stress) to achieve a separation of the planes of this amount will be $f/a^2 = kx/a^2$. This displacement is equivalent to a linear strain of amount $x/a$. Then we

Figure 52. Simple cubic array of atoms, (a) undeformed, (b) when a force $f$ is applied to each atom in plane III all the planes undergo an increased separation $x$ (c) a single force $f$ acts in a similar way on a single atom q.

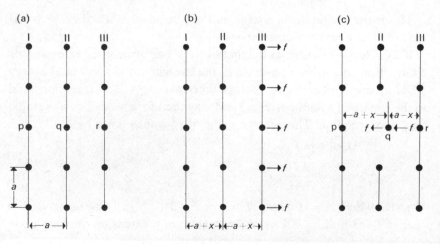

may write for Young's modulus

$$E = \frac{\text{tensile stress}}{\text{strain}} = \frac{kx/a^2}{x/a} = \frac{k}{a}. \tag{7.10}$$

## 7.2 Propagation of longitudinal waves along an elastic bar

We extend this to a very simple description of wave propagation. Our major (and weakest) assumption is that we may isolate the inter-action of one atom with its next neighbour. Suppose we hold planes I and III fixed (see figure 52($c$)) and displace atom q relative to atoms p and r by an amount $x$. Then p exerts an attractive force $kx$ on q and r exerts a repulsive force of the same magnitude. The resulting restoring force is $2kx$. If the atomic mass is $m$ the equation of motion is

$$m\ddot{x} + 2kx = 0.$$

This gives simple harmonic motion of period

$$T = 2\pi \left(\frac{m}{2k}\right)^{\frac{1}{2}}. \tag{7.11}$$

Suppose now we give an impulse to the free end of the bar and displace the row of atoms in plane I to the right (figure 52). These then act on the row of atoms in plane II which in turn are pushed to the right and so act on the atoms in plane III. In this way the impulse is propagated through the solid. The time taken for each plane to respond to the push of the neighbouring plane is of the order of one-quarter of the vibrational period $T$. Thus the impulse jumps from one plane to the next in a time

$$t \approx \frac{2\pi}{4} \left(\frac{m}{2k}\right)^{\frac{1}{2}} \approx \frac{2\pi}{5.7} \left(\frac{m}{k}\right)^{\frac{1}{2}}. \tag{7.12}$$

For our purpose we may take $2\pi/5.7$ to be unity. The disturbance thus travels a distance $a$ in time $(m/k)^{\frac{1}{2}}$. The velocity of propagation is therefore

$$v = \frac{a}{t} = a \left(\frac{k}{m}\right)^{\frac{1}{2}},$$

or

$$v = \left(\frac{a^3}{m} \frac{k}{a}\right)^{\frac{1}{2}}. \tag{7.13}$$

We recognize $m/a^3$ as the density $\rho$, and the quantity $k/a$ as Young's modulus $E$. Hence

$$v = \left(\frac{E}{\rho}\right)^{\frac{1}{2}}. \tag{7.14}$$

This is the same as the equation for the velocity of a longitudinal wave in a bar in terms of its macroscopic properties. It shows that the vibrating atom is the 'messenger' for wave propagation.

The frequency of vibration is

$$\nu = \frac{1}{T} = \frac{1}{2\pi}\left(\frac{2k}{m}\right)^{\frac{1}{2}} = \frac{\sqrt{2}}{2\pi a}\,v,$$

$$\nu = \frac{1}{2\pi a}\left(\frac{2E}{\rho}\right)^{\frac{1}{2}}. \tag{7.15}$$

For, say, iron we have $E = 2 \times 10^{11}\,\mathrm{N\,m^{-2}}$, $\rho = 7.8 \times 10^{3}\,\mathrm{kg\,m^{-2}}$, $a = 2.5 \times 10^{-10}\,\mathrm{m}$. Then $\nu \approx 5 \times 10^{12}$ vibrations per second. This is the right order of magnitude.

The main defect in this model is that the vibrations of the atoms are not isolated but are coupled. A rigorous treatment does not, however, substantially alter the nature of this analysis.

### 7.3 Bulk moduli

#### 7.3.1 *Bulk modulus of an ionic solid*

We consider here an ionic solid of the NaCl type in which the $\mathrm{Na^+}$ and $\mathrm{Cl^-}$ ions are distributed on a cubic lattice as shown in figure 53. We consider a single ion and its coulombic interaction with all the charged ions around it. As the figure shows there are:

6 neighbours of opposite sign at distance $x$

12 neighbours of same sign at distance $\sqrt{2}x$

8 neighbours of opposite sign at distance $\sqrt{3}x$, etc.

If each ion has charge $e$ (or $-e$) the potential energy of one ion in relation to all its neighbours is

$$\frac{-e^2}{x}\left(6 - \frac{12}{\sqrt{2}} + \frac{8}{\sqrt{3}}\cdots\right)\frac{1}{4\pi\varepsilon_0}$$

$$= \frac{-e^2}{x}(2.1\ldots)\frac{1}{4\pi\varepsilon_0}. \tag{7.16}$$

Proper summation to infinity shows that the correct value, known as the Madelung constant $A$, is 1.748, say 1.75, for the rock-salt structure.

Assuming a repulsive potential energy term between each ion pair of the form $A/x^9$, the summation of all the repulsion terms will give $B/x^9$, where $B$ is an appropriate constant. The potential energy of one ion allowing for both attraction and repulsion is

$$u = \left(-1.75\frac{e^2}{x} + \frac{B}{x^9}\right)\frac{1}{4\pi\varepsilon_0}. \qquad (7.17)$$

If there are $N$ positive and $N$ negative ions per mole the potential energy for a mole of NaCl is

$$U = \tfrac{1}{2}(2N_A u) = N_A\left(-1.75\frac{e^2}{x} + \frac{B}{x^9}\right)\frac{1}{4\pi\varepsilon_0}, \qquad (7.18)$$

where the $\frac{1}{2}$ is introduced so that we do not count each bond twice.

The equilibrium spacing $a$ occurs when $U$ is a minimum.

Putting $\partial U/\partial x$ at $x = a$ equal to zero we find that $B$ has the value $1.75e^2 a^8/9$. Equation (7.18) becomes

$$U = -1.75N_A e^2\left(\frac{1}{x} - \frac{a^8}{9x^9}\right)\frac{1}{4\pi\varepsilon_0}. \qquad (7.19)$$

The next step appears complicated but is basically an attempt to connect the internal energy with an elastic modulus, in this case the bulk modulus $K$. If we apply a hydrostatic pressure $P$ this produces a volume change $-dV$, and the work done by the external forces on the crystal is then $-P\,dV$. The ions are squeezed closer together and if the whole of this work goes solely in increasing that part of $U$ which arises from inter-atomic forces, we have $-P\,dV = dU$ or

$$P = -\frac{\partial U}{\partial V}. \qquad (7.20)$$

But

$$K = -\frac{dP}{dV/V} = \frac{\partial^2 U}{\partial V^2}V. \qquad (7.21)$$

Figure 53. The structure of the NaCl crystal. (*a*) General view; (*b*) 6 neighbours of opposite sign at distance $x$; (*c*) 12 neighbours of same sign at distance $\sqrt{2}x$; (*d*) 8 neighbours of opposite sign at distance $\sqrt{3}x$.

(a)   (b)   (c)   (d)

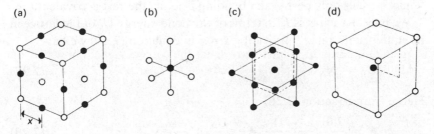

We need to express this in terms of the ionic spacing $x$. We have

$$\frac{\partial^2 U}{\partial V^2} = \frac{\partial}{\partial V}\left(\frac{\partial U}{\partial V}\right) = \frac{\partial}{\partial V}\left(\frac{\partial U}{\partial x} \times \frac{dx}{dV}\right) = \frac{\partial}{\partial x}\left(\frac{\partial U}{\partial x} \times \frac{dx}{dV}\right)\frac{dx}{dV}$$

$$= \frac{\partial^2 U}{\partial x^2}\left(\frac{dx}{dV}\right)^2 + \frac{\partial U}{\partial x}\left(\frac{d^2 x}{dV^2}\right) + \ldots \tag{7.22}$$

At the equilibrium separation $(x = a)$, $(\partial U/\partial x) = 0$; consequently the second term vanishes so that

$$K = V\frac{\partial^2 U}{\partial x^2}\left(\frac{dx}{dV}\right)^2. \tag{7.23}$$

We now relate $V$ to the atomic spacing $x$. For one mole

$$V = 2N_A x^3.$$

Therefore

$$\frac{dx}{dV} = \frac{1}{6N_A x^2}. \tag{7.24}$$

To obtain $K$ of equation (7.23) we differentiate equation (7.19) twice to obtain $\partial^2 U/\partial x^2$ and use equation (7.24) for $(dx/dV)^2$. We obtain

$$K = VN_A 1.75 e^2\left(\frac{-2}{x^3} + \frac{10a^8}{x^{11}}\right)\frac{1}{36N_A^2 x^4}\frac{1}{4\pi\varepsilon_0}. \tag{7.25}$$

At equilibrium when $x = a$ and $V = 2N_A a^3$ we find

$$K = \left(\frac{7.0}{9} \times \frac{e^2}{a^4}\right)\frac{1}{4\pi\varepsilon_0} = 0.78\frac{e^2}{a^4}\frac{1}{4\pi\varepsilon_0}. \tag{7.26}$$

For rock-salt, assuming complete charge separation on the ions, $e$ is the electronic charge, $1.6 \times 10^{-19}$ C. The equilibrium spacing $a = 2.82 \times 10^{-10}$ m. Substituting in equation (7.26) we obtain

$$K = 2.8 \times 10^{10}\,\text{N m}^{-2},$$

whereas the experimental value is about $2.4 \times 10^{10}\,\text{N m}^{-2}$. Taking into account the basic simplicity of the model this is surprisingly good agreement. It implies, of course, that the bonding is fully ionic. In most 'ionic' crystals only part of the bonding is ionic, the rest is covalent.

We may also express $K$ in terms of the bond energy $U_0$ of the lattice in its equilibrium state. Substituti$_1$g $x = a$ in equation (7.19) we have

$$U_0 = -N_A\frac{14.0}{9}\frac{e^2}{a}\frac{1}{4\pi\varepsilon_0}. \tag{7.27}$$

Hence from equation (7.26)

$$K = \left(\frac{7.0}{9} \times \frac{e^2}{a^4}\right)\frac{1}{4\pi\varepsilon_0} = \frac{-U_0}{2N_A a^3} = \frac{-U_0}{V}, \tag{7.28}$$

where $V$ is the volume of one mole. We shall see that a similar result is obtained for van der Waals solids.

It is interesting to note that, for a free NaCl molecule in the vapour state, the distance between the $Na^+$ and $Cl^-$ ions is $2.36 \times 10^{-10}$ m compared with $2.82 \times 10^{-10}$ m in the crystal lattice. This indicates the important part played by the surrounding ions of opposite sign in opening up the ionic spacing. It is relatively easy to make a rough quantitative estimate of this. For a single ionic pair in the free molecule we may write

$$u = \left(\frac{-e^2}{r} + \frac{A}{r^9}\right) \frac{1}{4\pi\varepsilon_0}. \tag{7.29}$$

The equilibrium spacing $r_0$ occurs when $\partial u/\partial r = 0$. This gives the value of $A = r_0^8 e^2/9$. Hence

$$u = -e^2 \left(\frac{1}{r} - \frac{r_0^8}{9r^9}\right) \frac{1}{4\pi\varepsilon_0}. \tag{7.30}$$

In the crystal the coulombic interaction of the ionic charges gives alternative positive and negative energies; as we saw above, the final amount for each ion is $-1.75 \, e^2/r$, where 1.75 is the Madelung constant. The repulsion term always involves a positive energy; however, the potential falls off so rapidly with distance that we need only consider the 6 nearest neighbours at distance $r$ for which the repulsion energy is $6A/4\pi\varepsilon_0 r^9$ or $6r_0^8 e^2/4\pi\varepsilon_0 9r^9$. The energy per ion is then

$$u = -e^2 \left(\frac{1.75}{r} - \frac{6r_0^8}{9r^9}\right) \frac{1}{4\pi\varepsilon_0}. \tag{7.31}$$

The minimum occurs when $\partial u/\partial r = 0$, i.e. when

$$\frac{1.75}{r^2} = \frac{6r_0^8}{r^{10}},$$

$$r^8 = 3.43r_0^8 \quad \text{or} \quad r = 1.17r_0. \tag{7.32}$$

If $r_0$ is taken as $2.36 \times 10^{-10}$ m this would give for $r$ a value of $2.76 \times 10^{-10}$ m which is close to the observed value. Even for a one-dimensional infinite chain molecule NaClNaClNaCl..., the ionic spacing is already increased from $2.36 \times 10^{-10}$ m to $2.69 \times 10^{-10}$ m.

### 7.3.2 *Bulk modulus of a van der Waals solid*

We consider a simple van der Waals solid. If it consists of single atoms or spherical or nearly spherical molecules it generally acquires a close-packed (f.c.c.) structure. If the molecules are long chain hydrocarbons the packing tends to be very nearly hexagonal with the chains lying parallel to one another. We shall discuss only the f.c.c. type of structure such as is found in solid argon, neon and krypton. For a pair of atoms at a distance $x$ apart the van der Waals attractive forces produce a potential energy proportional to $x^{-6}$; the repulsion produces a potential energy which appears to be proportional to $x^{-12}$, rather than to $x^{-9}$ as in the previous ionic model.

The potential energy between two atoms may be written

$$u = -\frac{A}{x^6} + \frac{B}{x^{12}}. \tag{7.33}$$

Each atom has 12 nearest neighbours distance $\sqrt{2}x$ apart and 6 neighbours $2x$ apart, etc. (see figure 54). The contribution from the attraction forces is

$$-\left(\frac{12A}{(\sqrt{2}x)^6} + 6\frac{A}{(2x)^6} + \cdots\right) = -\frac{A}{(\sqrt{2}x)^6}\left(12 + \frac{6}{8} + \cdots\right). \tag{7.34}$$

Although the convergence is very rapid the contribution from even more distant neighbours is not entirely negligible. A full summation shows that the quantity in the bracket reaches a limiting value of 14.45. This is the Madelung constant for this structure and shows that taking all neighbours into account the *effective* number of nearest neighbours is increased from 12 to 14.45. Thus the contribution from the attractive forces becomes

$$-\frac{A}{8 \times x^6} \times 14.45 \approx -\frac{1.8A}{x^6}. \tag{7.35}$$

Figure 54. The structure of an f.c.c. structure of a solid inert gas. Each atom has 12 neighbours at distances $\sqrt{2}x$ apart, 6 neighbours at distance $2x$ apart, etc.

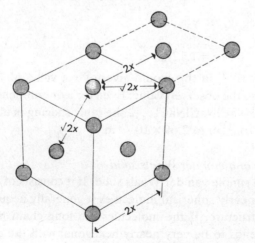

The repulsion term may be grouped in a single constant $C$ so that the resultant potential energy may be written

$$u = -\frac{1.8A}{x^6} + \frac{C}{x^{12}} \, . \tag{7.36}$$

For $N_A$ atoms the total energy will be $\frac{1}{2}N_A$ times this. As before we may differentiate $U$ to find the equilibrium condition when $x = a$.

$$U = -0.9 N_A A \left( \frac{1}{x^6} - \frac{1}{2} \frac{a^6}{x^{12}} \right). \tag{7.37}$$

At the equilibrium separation

$$U_0 = -0.45 \frac{N_A A}{a^6}. \tag{7.38}$$

The volume of the mole for this structure is

$$V = 2 N_A x^3, \tag{7.39}$$

so that as before

$$\frac{dx}{dV} = \frac{1}{6 N_A x^2}. \tag{7.40}$$

Repeating the previous procedure we find for the bulk modulus

$$K = \frac{1.8A}{a^9}. \tag{7.41}$$

As indicated in equation (7.35), for the attractive part of the potential (which falls off as $x^{-6}$) the Madelung constant is 14.45, i.e. the effective number of nearest neighbours is increased from 12 to 14.45. On the other hand, for the repulsive part where the energy falls off as $x^{-12}$ we may ignore everything beyond second-nearest neighbours. This is not only because the numerical quantities are so small but because the repulsion arises from the overlap of orbitals so that it is physically unreasonable to think of repulsive interactions between more distant atoms. This gives an effective Madelung constant of 12.09. (Textbooks which sum over all neighbours give a value 12.13). Thus the potential energy due to the attractive part gains from distant neighbours more than the energy due to the repulsive part. As a result the atoms are drawn closer together: and the atomic separation is slightly smaller in the solid crystal than in the simple diatom by about 3 per cent. This should be contrasted with the behaviour of an ionic solid such as NaCl.

As a first approximation we may assume that the nearest-neighbour distance of atoms in the crystal $\sqrt{2}x$ is the same as the equilibrium separation $r_0$ for a single pair of atoms. We may thus re-write equation (7.41) as

$$K = \frac{1.8}{a^3} \times \frac{8A}{r_0^6}. \tag{7.42}$$

The second factor in equation (7.42) has a very simple significance. By differentiating equation (7.33) we may determine $B$ in terms of the equilibrium $r_0$ of one atom in relation to its neighbour. We shall then find the potential energy $\Delta\varepsilon$ of an isolated pair of atoms is simply $-A/2r_0^6$. Consequently from equation (7.42)

$$K = -\frac{28.8\Delta\varepsilon}{a^3}. \tag{7.43}$$

For the inert gases and for the smaller molecules such as $H_2$ and $N_2$, $\Delta\varepsilon$ is of order $(5 \text{ to } 25) \times 10^{-22}$ J, and $a$ is about $3 \times 10^{-10}$ m in the solid state. We obtain

$$K = (4 \text{ to } 20) \times 10^8 \, N \, m^{-2}. \tag{7.44}$$

This is about one hundred times smaller than for ionic solids.

We may also express $K$ in terms of $U_0$. Combining equations (7.38) and (7.41) to eliminate $A$ we have

$$K = -4\frac{U_0}{N_A a^3} = -8\frac{U_0}{V}. \tag{7.45}$$

This is similar to the result for ionic crystals except that the numerical factor is appreciably larger.

### 7.3.3 Generalized relation between bulk modulus $K$ and lattice energy $U_0$

The previous sections dealing with ionic solids and van der Waals solids may now be put into a much more concise and general framework. Suppose the potential energy between constituent atoms or ions is of the form

$$u = -\frac{A}{x^n} + \frac{B}{x^m}. \tag{7.46}$$

Then the lattice energy can be written

$$U = -C\left(\frac{1}{nx^n} - \frac{a^{m-n}}{mx^m}\right), \tag{7.47}$$

where $C$ is a suitable constant including the Madelung constant and other parameters and $a$ is the equilibrium spacing. We see that this is the correct form for $U$ since if we differentiate equation (7.47) with respect to

$x$, we find that $\partial U/\partial x = 0$ at $x = a$ confirming that the energy is a minimum at the equilibrium spacing. Thus

$$\frac{\partial U}{\partial x} - C\left(\frac{1}{x^{n+1}} - \frac{a^{m-n}}{x^{m+1}}\right) \tag{7.48}$$

and

$$\frac{\partial^2 U}{\partial x^2} = -C\left(\frac{n+1}{x^{n+2}} - \frac{(m+1)a^{m-n}}{x^{m+2}}\right).$$

Consequently at $x = a$

$$\frac{\partial^2 U}{\partial x^2} = -C\left(\frac{n-m}{a^{n+2}}\right). \tag{7.49}$$

For any regular crystalline array the molar volume can be written

$$V = zN_A x^3,$$

where $z$ is a constant determined by the crystal structure. Then

$$\frac{dx}{dV} = \frac{1}{3zN_A x^2}.$$

Consequently at equilibrium

$$\left(\frac{dx}{dV}\right)^2_{x=a} = \left(\frac{1}{9(zN_A x^2)^2}\right)_{x=a} = \frac{a^2}{9V^2}. \tag{7.50}$$

Hence

$$K = V\left(\frac{\partial^2 U}{\partial x^2}\right)\left(\frac{dx}{dV}\right)^2$$

$$= -VC\left(\frac{n-m}{a^{n+2}}\right)\left(\frac{a^2}{9V^2}\right) = C\left(\frac{m-n}{a^n}\right)\frac{1}{9V}. \tag{7.51}$$

But from equation (7.47) the equilibrium lattice energy $U_0$ at $x = a$ is

$$U_0 = -C\left(\frac{1}{na^n} - \frac{1}{ma^n}\right) = -C\left(\frac{m-n}{mna^n}\right). \tag{7.52}$$

Thus equation (7.51) becomes

$$K = -\left(\frac{mn}{9}\right)\frac{U_0}{V}. \tag{7.53}$$

For rock-salt, $n = 1$; $m \approx 9$; the factor is unity. For a van der Waals solid, $n = 6$; $m = 12$; the factor is 8.

### 7.3.4 *Bond energy, or lattice energy, and heat of sublimation*

Before going on to discuss the bulk modulus of metals we make a few general comments on the bond (or lattice) energy and the heat of sublimation.

The total bond energy $U_0$ of a lattice (the lattice energy) is a fundamental quantity deriving from the forces between the constituent atoms and ions. It is not easily open to direct measurement. The nearest physical quantity is the heat of sublimation $L_S$, which in practice is a little greater than the sum of the latent heat of fusion $L_F$ and the latent heat of vaporization $L_V$. We may easily form an estimate of $L_S$ for an ionic crystal. When sodium chloride is evaporated, the process does not involve pulling the positive and negative ions apart to infinity; this would be the true lattice energy. Vaporization generally produces individual NaCl molecules. Thus $L_S$ is less than $U_0$ by the value of the interaction energy between Na$^+$ and Cl$^-$ in the individual NaCl molecule. From equation (7.31) the lattice energy per ion in the solid state is obtained by putting $r = 1.17 r_0$, where $r$ is the ionic spacing in the crystal, $r_0$ the spacing in the free NaCl molecule. The value is

$$u_{\text{lattice}} = -\frac{e^2}{r_0} \frac{1.33}{4\pi\varepsilon_0},$$

and for one mole

$$U_0 = -\frac{N_A e^2}{r_0} \frac{1.33}{4\pi\varepsilon_0}.$$

The energy per ion pair in the free molecule is obtained from equation (7.30) by putting $r = r_0$. We obtain for the free NaCl molecule

$$u_{\text{molecular}} = -\frac{e^2}{r_0} \left(\frac{8}{9}\right) \frac{1}{4\pi\varepsilon_0},$$

and for one mole

$$U_{\text{molecular}} = -\frac{N_A e^2}{r_0} \left(\frac{8}{9}\right) \frac{1}{4\pi\varepsilon_0}.$$

The molar heat of sublimation is therefore

$$L_S = U_0 - U_{\text{molecular}} = \frac{N_A e^2}{r_0} \frac{0.44}{4\pi\varepsilon_0} \approx -\frac{1}{3} U_0. \tag{7.54}$$

The latent heat of sublimation, which we can measure with a fair degree of accuracy, is thus about one-third of the lattice energy. Tables of values support this conclusion for ionic solids.

With a van der Waals solid, sublimation merely separates all the molecules (or atoms) as individual molecules (or atoms). The heat of sublimation is thus a fairly accurate measure of the bond energy. We may easily form an estimate of this from the above analysis.

From equation (7.45) we have

$$U_0 = -\frac{KV}{8}$$

and from equation (7.43), this becomes

$$U_0 = \frac{3.6\Delta\varepsilon V}{a^3}. \tag{7.55}$$

But from equation (7.39) we see that when the crystal is at its equilibrium separation $(x = a)$, $V = 2N_A a^3$. Substituting in equation (7.55) we have

$$U_0 = 7.2 N_A \Delta\varepsilon. \tag{7.56}$$

This is the same as the value obtained in equation (6.2) except that the numerical factor is slightly different. This arises from the fact that in the present model we have included the effect of more distant neighbours.

Metals present a very special problem. They consist of metal ions in a sea of free electrons. It is not possible to describe the forces between the ions and electrons in terms of a single power law. Consequently the analysis we have given above is not possible. A reasonably good estimate of the bond energy $U_0$ can be made, but the treatment is very difficult and involves concepts which we shall not discuss in this book. Even if we knew the bond energy it would not correspond to the latent heat of sublimation. The bond energy would correspond to the work done in separating all the ions and all the free electrons to infinity; whereas in the vapour state the metal exists essentially as free atoms. Thus $U_0 = L_S +$ ionization energy of the metal atoms. A similar issue applies to covalent solids and partly explains why $\gamma_{calc} > \gamma_{obs}$ (see p. 156).

### 7.3.5 *Bulk modulus of metals*

For the reasons given above it is not possible to derive the elastic properties of a metal using the simple type of analysis that we have applied to ionic and van der Waals solids. However, we may form a reasonable estimate of the bulk modulus by noting that, in general,

$$K = -c\frac{U_0}{V}, \tag{7.57}$$

where $U_0$ is the total bond energy per mole, $V$ the atomic (or molecular) volume and $c$ a constant depending on the law of forces. For an ionic crystal $c = 1$; for a van der Waals solid it is 8. For metals we can only guess at a value of $c$. We also have to recognize that we have no simple way of calculating $U_0$, but we can use the sublimation energy $L_S$ (which can be determined experimentally) as some measure of $U_0$. As a matter of interest we shall try the relation

$$K = 3\frac{L_S}{V},$$ (7.58)

where 3 is purely an empirical factor. This represents an attempt to allow both for the unknown law of force between metal ions in the lattice and for the discrepancy between the bond energy and the latent heat of sublimation. For $L_S$ we have simply added the latent heats of fusion and of vaporization,† and have ignored any specific heat capacity terms. The results are given in Table 24. The agreement is surprisingly good in view of the over-simplifying assumptions made.

## 7.4 Elastic properties of rubber

The elastic properties of rubber are very different from those of most other solids such as ionic solids, covalent solids and metals. In tension or shear the modulus is 1000 to 10 000 times smaller. On the other hand Poisson's ratio is almost $\frac{1}{2}$, so the bulk modulus is relatively high; it approaches that of a liquid. The two most striking elastic features are (*a*) rubber has enormous extensibility – up to 400 per cent compared with perhaps 1 per cent for ordinary solids, (*b*) the elastic modulus increases with increasing temperature. This is the only solid known which shows such an effect. We recognize that these two properties are very similar to the behaviour of a gas in compression.

Since rubber is known to consist of long chain organic molecules we realize at once that its enormous extensibility and low modulus cannot arise from the stretching or bending of the C–C bonds, since these are

Table 24. *Latent heat of sublimation $L_S$ and bulk modulus $K$ of some metals*

| Metal | $L_S$ mol$^{-1}$ (Kcal) | $(J \times 10^3)$ | Atomic vol $V$ $(10^{-6}\,m^3)$ | $K \times 10^{10}$ (N m$^{-3}$) Calc. | Observed |
|-------|------|------|------|------|------|
| K  | 22  | 92  | 45.5 | 0.6 | 0.4  |
| Mg | 36  | 152 | 14.0 | 3.3 | 3.6  |
| Pb | 47  | 198 | 17.7 | 3.4 | 4.6  |
| Al | 77  | 322 | 10   | 9.7 | 7.5  |
| Cu | 82  | 345 | 7    | 15  | 13.8 |
| Fe | 95  | 400 | 7.1  | 17  | 17.8 |
| Mo | 159 | 670 | 9.4  | 21  | 26.1 |
| W  | 203 | 850 | 9.5  | 27  | 31.1 |

† Taken from C. J. Smithells, *Metals Reference Book*, 3rd edn (Butterworth, 1962).

extremely strong and difficult to deform. The rubbery behaviour is due to easy *rotation* of the carbon atoms about each C–C bond. As a result the chains are coiled up in higgledy-piggledy fashion and the result of applying a tension is to pull the chains into partial alignment. Thermal energy attempts to increase the disorder in the chain and therefore resists the aligning action of the tensile force. Consequently the modulus in tension increases with temperature.

If the molecular chains were completely separate the material would be a very viscous liquid and indeed natural latex is such a liquid. In order to become a rubbery solid the molecular chains must be cross-linked at a relatively small number of points. We now consider briefly the properties of a typical segment lying between two cross-links. Suppose the C–C bond has a length $\lambda$ and there are $N$ such bonds in a segment; because, as mentioned above, *rotation* about the C–C bond, in contrast to stretching or bending, is extremely easy, as a first approximation we may ignore any stiffness which arises from this factor. The segment then consists of $N$ freely pivoted bonds. We may then ask: what is the most probable length of a segment of $N$ units each of length $\lambda$?

Assuming that each link is completely free to take up any position relative to its neighbour we recognize this as the problem of a 'random walk' in three dimensions or a 'random flight.†

We indicate briefly here how this may be calculated. Consider first a random walk in one dimension. Suppose a man takes $N$ steps, each of length $\lambda$, along a line; each step may be forwards or backwards. What is the probability that he will have travelled a distance $x$ from his starting point? If out of $N$ steps $A$ are forwards and $B$ are backwards

$$A + B = N$$

and

$$A\lambda - B\lambda = x. \tag{7.59}$$

Hence

$$A = \tfrac{1}{2}\left(N + \frac{x}{\lambda}\right),$$

$$B = \tfrac{1}{2}\left(N - \frac{x}{\lambda}\right). \tag{7.60}$$

---

† This ignores the requirement of fixed bond-angles between the carbon atoms. If this is taken into account, and rotations about the bonds are permitted, the value of $\beta$ in equation (7.66) is merely changed by a small numerical factor. A more serious conceptual difficulty is that in the random walk any given step may be covered many times, whereas in a real molecule the location of a molecular link can only be occupied once.

Any way of taking $N$ steps such that $A$ are forwards and $B$ backwards will give the resultant distance $x$. The number of ways $W$ in which this can be achieved is

$$W = \frac{N!}{A!\,B!}. \tag{7.61}$$

For simplicity assume $N$ to be even so that $A$ and $B$ (from equation (7.60)) are integers. We use the Stirling approximation $N! = (N/e)^N$ and take $x \ll N\lambda$ (in the rubber molecule this is equivalent to assuming that the distance between the two ends of the segment is very much less than the fully extended length of the segment). Then

$$\ln W = N \ln 2 - \frac{x^2}{2N\lambda^2}$$

or

$$W = B \exp(-\beta_1^2 x^2), \tag{7.62}$$

where $\beta_1^2 = 1/2N\lambda^2$. The probability $P(x)$ of the distance $x$ occurring is directly proportional to $W$ so that

$$P(x) = C \exp(-\beta_1^2 x^2). \tag{7.63}$$

The constant $C$ may be determined by observing that

$$\sum_{x=-N\lambda}^{x=N\lambda} P(x) = 1 \approx \int_{-\infty}^{\infty} P(x)\,\mathrm{d}x. \tag{7.64}$$

This gives $C = \beta_1/\pi^{\frac{1}{2}}$.

We now turn to the three-dimensional random walk (figure 55($a$)). If the coordinates of the starting point are $(0, 0, 0)$ and those of the final point are $(x, y, z)$ on orthogonal axes, the probability of finding the final position is

$$P(x, y, z) = \frac{\beta^3}{\pi^{\frac{3}{2}}} \exp[-\beta^2(x^2 + y^2 + z^2)]. \tag{7.65}$$

Normalizing for three dimensions it turns out that the $\beta^2$ in equation (7.65) is 3 times $\beta_1$ for the one-dimensional random walk, i.e.

$$\beta^2 = \frac{3}{2} \frac{1}{N\lambda^2}. \tag{7.66}$$

We are only interested in the exponential factor. We may therefore write for the most probable length of a segment

$$P(r) = A \exp(-\beta^2 r^2). \tag{7.67}$$

If one end of the segment is fixed, the probability of finding the other end in an element of volume $d\tau$ is

$$P \, d\tau = A \exp(-\beta^2 r^2) \, d\tau. \qquad (7.68)$$

We see that the probability density $P$ is a maximum when $r = 0$, i.e. the segment makes its greatest effort to have zero distance between its ends. This is to be expected since if all steps in the random walk are equally probable a large number will contain as many forward as backward ones. This does not mean that the most probable distance between chain ends is zero. If we are interested only in the length $r$ and not the direction then (as in our treatment of molecular velocities in a gas) we may consider one end of the segment fixed and determine the number of the other ends lying in

Figure 55. (*a*) Probability density $P$ as function of distance between the ends of a long chain assuming a three-dimensional 'random walk', (*b*) most probable distance $r$ between the ends of a chain with $N$ segments each of length $\lambda$ occurs for $r_m = 1/\beta = \lambda(\frac{2}{3}N)^{\frac{1}{2}}$, (*c*) random arrangement of a chain containing 1000 segments, each segment set at the appropriate bond angle, but being given a random choice of 6 equally-spaced angular positions; this is a reasonably close approximation to complete randomness (from L. R. G. Treloar, *Physics of Rubber Elasticity*, 2nd edn (Oxford University Press, 1958)).

(a)

(c)

a shell of volume $4\pi r^2\,dr$. Then the number $\Pi$ with lengths between $r$ and $dr$ is

$$\Pi\,dr = A4\pi r^2 \exp(-\beta^2 r^2)\,dr.$$

The result is shown in figure 55($b$); the maximum occurs for a value of $r$ given by

$$r_m = \frac{1}{\beta} = \lambda(\tfrac{2}{3}N)^{\frac{1}{2}}. \tag{7.69}$$

A simple analogy is to consider the concentration of bees around a drop of honey. The greatest density occurs near the honey but the greatest number of bees may well be found between 5 and 6 cm from the drop, rather than in the first centimetre.

Reverting to equation (7.67) the probability density is $P$ and this is related to the (configurational) entropy of the segment by the relation

$$S = k \ln P = \text{constant} - k\beta^2 r^2. \tag{7.70}$$

If we apply a tensile force $f$ to one end of the segment and extend it in the direction of $r$ by an amount $dr$, the external work done is $f\,dr$. Assuming that this involves no volume change we may equate it to the change in the Helmholtz free energy (equation (2.23)).

$$f\,dr = dA = d(U - TS). \tag{7.71}$$

For an isothermal extension

$$f = \frac{\partial U}{\partial r} - T\frac{\partial S}{\partial r}$$

$$= \frac{\partial U}{\partial r} + 2kT\beta^2 r. \tag{7.72}$$

The term $\partial U/\partial r$ refers to the change in *internal* energy with extension, that is, with uncoiling of the segment by rotation about the C–C bond. As there is very little energy change (in our model no change at all) in this we may neglect this term and write

$$f = 2kT\beta^2 r = \frac{2RT\beta^2 r}{N_A}. \tag{7.73}$$

We see at once that ($a$) $f$ is proportional to $r$ so that the segment has Hookean properties, ($b$) $f$ is proportional to $T$, ($c$) $f$ is proportional to $\beta^2$, i.e. to $1/N$: the smaller the number of bonds in the segment, the greater its resistance to elastic extension.

So far we have only discussed the force that the segment exerts in resisting extension; whatever its length it is always attempting to contract.

In bulk rubber which consists of a network of segments and chains it turns out that even if the rubber is subjected to uniaxial compression the overall effect, because of the large Poisson's ratio, is to extend the chains. The specimen therefore resists both extension and compression in a manner resembling the behaviour of a single segment discussed above.

We see that the elasticity of rubber arises from the entropy or randomness of the chain segments. It is for this reason that it resembles the elasticity of an ideal gas in compression where the pressure may be considered as arising from the reduction in entropy associated with a reduction in volume. Indeed one notices the presence of the 'gas constant' $R$ as a fundamental term in the expression for rubber elasticity in equation (7.73). In view of this one might well designate $R$ as the 'rubber constant' in 'ideal rubber' elasticity.

# 8

# *The strength properties of solids*

## 8.1 **Deformation properties**
### 8.1.1 *Ductile properties*

The strength properties of solids are most simply illustrated by considering the behaviour of a homogeneous specimen of uniform cross-section subjected to uniaxial tension (figure 56($a$)). If we plot the true stress $\sigma$ against the linear strain $\varepsilon$ (i.e. the fractional increase in length) we may obtain a curve as illustrated in figure 56($b$). The portion OA represents elastic deformation. The strain is proportional to the stress and the deformation is reversible. If the material is ductile, elastic deformation will proceed until at some critical stress $Y$ the onset of permanent

Figure 56. ($a$) Tensile specimen and ($b$) typical stress–strain curve, showing elastic deformation along OA, plastic yielding at $Y$ and work-hardening along YZ. A brittle solid fails at a tensile stress $S$.

or plastic deformation occurs. If we continue along the plastic curve there is generally an increase in yield stress with deformation. This is known as work-hardening. If at some point B we reduce the stress the material recovers elastically along BO', where BO' is very nearly parallel to OA. The displacement OO' is the permanent plastic extension produced in the specimen. On reapplying the stress the deformation follows the curve O'BZ.

Experiments show that under simple uniaxial compression a cylinder will first deform elastically and then yield plastically at the same compressive stress $Y$.

Some insight into the yield criterion is provided by considering the effect of hydrostatic pressure, i.e. a stress situation in which the comprehensive stress on the material is the same in all directions. If we subject the specimen to a hydrostatic pressure $P$ we find that plastic flow does not occur even if $P$ exceeds $Y$. We must still apply a uniaxial stress (either in tension or compression) and its magnitude if plastic flow is to occur is still $Y$. Now analysis shows that the only part of a stress field which is unaffected by hydrostatic pressure is a shear stress. We conclude that plastic flow is associated with a critical shear stress. This is fully supported by microscopic studies which show that plastic deformation is always accompanied by slip of atomic planes over one another.

Figure 57. Shear stresses produced by tensile stresses: ($a$) vertical stress $\sigma = F/A$, ($b$) horizontal stress $\sigma$, ($c$) two-dimensional hydrostatic stress $\sigma$; this produces a resultant zero shear stress.

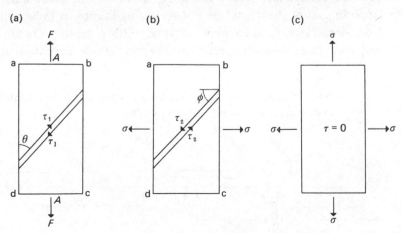

### 8.1.2 *Shear stress*

Consider a rectangular bar of uniform cross-sectional area $A$ (figure 57($a$)). Suppose we apply a tensile force $F$, so that the stress is $\sigma = F/A$. Consider a thin slice of material making an angle $\theta$ to the direction of $F$. On one side of the slice there is a force $F \cos \theta$, on the other side a force of equal magnitude in the opposite direction. These forces constitute a shear. The surface area of the slice is $A/\sin \theta$ so that the shear stress is given by

$$\tau_1 = \frac{F \cos \theta}{A/\sin \theta} = \frac{F}{A} \cos \theta \sin \theta$$

$$= \sigma \cos \theta \sin \theta. \qquad (8.1)$$

The maximum occurs for $\theta = 45°$ and has a value $\tau = \sigma/2$.

If we apply a tensile stress $\sigma$ to the faces bc, da, the shear stress on the slice is

$$\tau_2 = \sigma \cos \phi \sin \phi$$

$$= \sigma \sin \theta \cos \theta. \qquad (8.2)$$

This stress is equal to $\tau_1$ and in the opposite direction. If the two systems of tensile stresses $\sigma$ are superposed (to constitute a two-dimensional hydrostatic tension), the two shear-stress systems completely annul one another. This is the basis for the statement that hydrostatic stresses do not in any way change the existing shear stresses in a system.

Finally we note that the conventional method of representing a shear stress by two equal and opposite parallel stresses $\tau$ is misleading (figure 58($a$)). Such a stress system would produce a couple which would produce continuous rotation of the stressed element. There must be an opposing set of shear stresses $s$ if static equilibrium is to be achieved. Consider a rectangular element length $x$, width $y$, depth $z$. On the upper and lower faces we apply shear stresses $\tau$; on the side faces shear stresses

Figure 58. Sketch showing that a shear stress must always involve a pair of equal orthogonal shear stresses.

(a)            (b)            (c)

s. For equilibrium take moments about O. Shear stress $\tau$ implies a shear force $\tau \times xy$; its couple is $\tau \times xy \times z$. Similarly shear stress $s$ implies a couple $s \times yz \times x$.

Then

$$\tau xy \times z = s \times yz \times x,$$

$$\tau = s. \tag{8.3}$$

We see that a shear stress always involves a pair of equal orthogonal stresses. In practice, of course, one is often unaware of this. For example in shearing a long thin slice as in figure 58($c$), the horizontal shear force $F_1 = \tau A_1$ may be very large and the vertical shearing force $F_2 = \tau A_2$ may be vanishingly small if $A_2$ is very small, although the value of the shear stress $\tau$ must be the same in both cases.

A tensile stress $Y$ produces a maximum shear stress $Y/2$ at 45° to the direction of $Y$. (The same applies to a uniaxial compressive stress $Y$.) For an isotropic material, shear will therefore occur in slip directions at 45° to the direction of the applied stress. If the material is not isotropic shear may occur more easily in some directions than others; if the shear stress is exceeded in this direction slip will occur in these more favourably orientated directions. This is often observed with single crystals, and is shown schematically for a simple shear stress in figure 59.

We may now explain the behaviour shown in figure 56($b$). When the stress is first applied the atoms are displaced from their equilibrium positions; the resistance to deformation is determined by the interatomic forces. When the tensile stress reaches a critical value the shear stress is sufficient to produce slip along an appropriate plane and plastic yielding occurs. However, the whole stress must still be supported by the displaced atoms. Consequently when the stress is removed there is elastic recovery and the modulus, which arises from the interatomic forces, is essentially the same as originally.

Figure 59. The effect of a shear stress on the yielding of a single crystal ($a$) when the shear plane is parallel to the shear, ($b$) when it is inclined.

(a)  (b)

### 8.1.3 *Indentation hardness of ductile solids*

Engineers and metallurgists often wish to determine the strength properties of their materials without going to the trouble of preparing tensile specimens and carrying out full-scale tensile tests. A very convenient way of doing this is to measure the indentation hardness. A very hard indenter (a hard steel sphere in the Brinell test, a diamond pyramid in the Vickers test) is pressed under a load $W$ into the surface of the material to form a plastic indentation. When the indenter is removed the diameter of the indentation is measured and its area $A$ determined. The mean pressure over the indentation is then

$$p = \frac{W}{A}. \tag{8.4}$$

In the industrial test procedure $A$ is the surface area of the indentation; from the point of view of making $p$ physically meaningful, the projected area is more appropriate. The difference, however, is generally small. A study of the stress situation around the indenter shows that almost two-thirds of $p$ is in the form of a hydrostatic component, and therefore plays no part in producing plastic flow. Consequently only one-third of $p$ is active in producing the indentation. Thus as a first approximation.

$$p = 3Y, \tag{8.5}$$

where $Y$ is the uniaxial yield stress of the material. This relation is well substantiated in practice.

Further, if the material does not appreciably work-harden after yielding in tension its yield stress $Y$ will be very nearly the maximum stress the material can support before it pulls apart, i.e. its ultimate tensile strength (u.t.s.). The latter is therefore about one-third the indentation hardness, if they are measured in the same units. Since $p$ is normally measured in kg $mm^{-2}$ and the u.t.s. in the older literature was measured in ton in.$^{-2}$, the conversion factor must be divided by 1.58. Consequently

$$\text{u.t.s. (ton in.}^{-2}) = \frac{0.33}{1.58} p = 0.21p \text{ (kg mm}^{-2}). \tag{8.6}$$

This conversion ratio is widely used for polycrystalline homogeneous materials and is reliable to within a few per cent.

Typical values for the indentation hardness in kg $mm^{-2}$ are 0.25 for krypton at 50 K, 0.8 for ice at 260 K, 1 for indium, 4 for lead, 40 for polycrystalline copper, 120 for mild steel, 900 for ball-bearing steel, 2000 for sapphire and over 10 000 for diamond at room temperature.

### 8.1.4 *Calculation of critical shear stress for single crystals*

We now consider the shear stress at which two neighbouring planes in a single crystal can be caused to slide over one another. Consider the arrangement in figure 60 for a typical f.c.c. crystal and apply a shear stress $\tau$ to the planes X and Y. If $\theta$ is the strain produced and $G$ is the rigidity or shear modulus

$$\tau = G\theta.$$

Atom B in its initial site in the lattice is in a position of minimum potential energy. As it is displaced the potential energy increases until a position of unstable equilibrium is reached at B' – the potential energy curve at this point is horizontal but is a maximum. The potential energy curve thus has the form shown in figure 60(*b*). The force curve is found by differentiating the potential energy curve: it is drawn in figure 60(*c*) and shows that the maximum force occurs when B has reached some intermediate position B'' between B and B'. The angle of strain at this point is about $\frac{1}{4}$. If $G$

Figure 60. Shearing of a perfect crystal. (*a*) Neighbouring planes of atoms at X and Y, (*b*) the potential energy curve as the top plane is slid over the bottom, (*c*) the equivalent force–displacement diagram, (*d*) a more realistic form of the force–displacement curve.

remained constant we could say that B is sheared over to B″ for a shear stress of magnitude

$$\tau_m \approx \frac{G}{4}. \tag{8.7}$$

The lattice could not resist a greater stress than this so that plane X would glide over plane Y for this value of $\tau$.

This is an overestimate since it assumes that $G$ is a constant and has the same value as for small strains. This is rather unrealistic since for larger distortions we expect the modulus to decrease. We may form a better estimate of $\tau_m$ by assuming that the shear stress is a sine function of the displacement. It must clearly be zero for displacements $x = 0$, $x = a/2$ and $x = a$, where $a$ is the atomic spacing. We may thus use a function of the form

$$\tau = A_0 \sin\left(\frac{2\pi}{a}x\right), \tag{8.8}$$

where $A_0$ is a suitable constant. We may determine $A_0$ by specifying that the initial slope near $x = 0$, where the strains are small, corresponds to the shear modulus $G$ of the material. For small $x$

$$\tau = A_0 \sin\left(\frac{2\pi}{a}x\right) = A_0 \frac{2\pi}{a}x = G\theta, \tag{8.9}$$

when $\theta$ is the shear angle $x/a$. Consequently

$$A_0 \frac{2\pi}{a}x = G\frac{x}{a} \quad \text{or} \quad A_0 = \frac{G}{2\pi}. \tag{8.10}$$

The shear stress reaches its maximum when the displacement is midway between B and B′; here $\sin(2\pi x/a)$ is unity so that

$$\tau_m = A_0 = \frac{G}{2\pi}. \tag{8.11}$$

A stress of this value will carry atom B over to position B″ and then for a slight increase in $\tau$ it will flick over to B′ and then to C. Thus slip along the atomic plane will take place for a shear stress of about $G/6$. A more realistic study of the atomic force field suggests a curve like that shown in figure 60($d$). This reduces the critical shear stress $\tau_m$ even further but no amount of adjustment can reduce it to a value less than about $G/30$.

We conclude that the maximum shear stress $\tau_m$ the lattice can withstand is of the order of $\frac{1}{10} G$ to $\frac{1}{30} G$. Before discussing the relation of $(\tau_m)$ theoretical to $(\tau_m)$ experimental we may ask a very simple but pertinent question. Once we have dragged atom B to the top of the potential hill at

B' (figure 60(*b*)) why does it not glide over the rest of the potential hills until planes X and Y have completely slid off one another? Is not the position analogous to a 'frictionless' helter-skelter, where once the carriage has reached its highest point it can proceed indefinitely over all hills which do not exceed that height? The simple answer is that such a model would be appropriate if the atoms in row Y were absolutely rigid in space. Because they are themselves part of an elastic system the behaviour is much more like the interaction of two (frictionless) combs, the teeth of one being dragged through those of the other (figure 61(*a*)). Up to a certain tangential stress both set of teeth are distorted reversibly (figure 61(*b*)); beyond this point the teeth from the upper comb escape from interaction with the lower comb and flick over into their new equilibrium position; whilst the teeth of the lower comb flick back into their original position (figure 61(*c*)). The whole of the distortional energy has disappeared as vibration of the teeth. This has a close analogue in the slip of crystals. Practically the whole of the work of plastic deformation is dissipated as vibrational energy in the lattice, i.e. as heat. Not more than a few per cent of the energy is retained as strain energy in the lattice. This has an interesting corollary in an unexpected area. The friction of metals is largely due to adhesion, shearing and deformation within and around the regions of real contact; it is, in fact, a process which is often dominated by plastic flow around the contact zones. The plastic work virtually all appears as heat. This is the reason for Joule's observation that in frictional heating there is quantitative agreement between the work done against friction and the heat liberated. If large amounts of energy could be stored in the lattice this equivalence would not be observed.

Figure 61. The displacement of the teeth of two sets of engaging combs (*a*) initially, (*b*) after a small stress has been applied, (*c*) after the stress has been sufficient to cause slip: the elastic strain energy in the teeth is lost by vibration.

(a)   (b)   (c)

## 8.2 Dislocations

### 8.2.1 *Observed shear stress, need for dislocations*

Experiments on pure single crystals show that minute shear stresses are sufficient to produce permanent, i.e. plastic, deformation. In general ($\tau_m$) observed is of the order of one thousand times smaller than ($\tau_m$) theoretical. The theoretical value cannot be wrong to this extent. This discrepancy has led to a search for sources of weakness in the lattice, such as flaws or imperfections: the simplest type is that known as the edge dislocation shown in figure 62($a$). It involves, essentially, the insertion of an extra half-plane of atoms AB into the lattice. If a shear stress is applied as shown, atom B flicks over and joins with atom F for a very small value of $\tau$. Then C joins up with G and D with H so that for a very small stress we have produced slip by one atomic spacing DE (figure 62($b$)). In effect what we are saying here is that the theoretical value of $\tau_m$ assumes that slip occurs simultaneously across the whole plane KE; in the dislocation model slip is able to occur one row of atoms at a time and this leads to an enormous reduction in shear stress.

We can already see that there are two basic problems associated with the above model. First, how are the dislocations originally produced in the crystal? Secondly when the edge dislocation has slipped out of the crystal as in figure 62($b$), the crystal has become perfect so that it has its ideal theoretical strength. As this is not generally observed there must be

Figure 62. ($a$) An edge dislocation, ($b$) the displacements which occur when a shear stress is applied and the dislocation at B moves out to E.

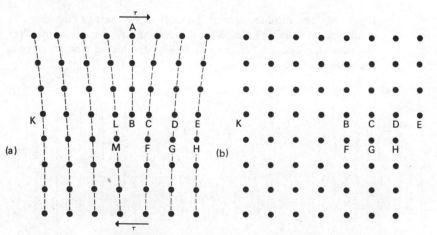

some process for the generation of further dislocations whilst slip is occurring. Both these questions are part of a very large subject which we shall not pursue further in detail. We mention, however, a few further points of interest.

### 8.2.2 *Direct experimental evidence for dislocations*

The first and most direct evidence for the existence of dislocations is due to Dr. J. W. Menter, who studied, with an electron microscope, very thin specimens of platinum phthalocyanine. This is a metallo-organic compound which forms well-defined crystals in which the platinum atoms lie on crystal planes 12 Å apart. In the electron microscope the organic part of the crystal scarcely absorbs electrons whilst the platinum atoms strongly absorb. In transmission, therefore, a microscope with a resolution of say 10 Å (1 nm) is able to resolve the planes of the platinum atoms. Typical electron micrographs reveal the presence of a defect strongly resembling a classical edge dislocation.†

In pure metals the atomic spacing is of the order 3 Å (0.3 nm) and it is only in recent years that electron microscopes have been developed with sufficient resolution to 'see' dislocations directly in metals. However, even with poorer resolution it is possible to identify certain features in the microscope as dislocations. This has led to a very extensive application of electron microscopy and diffraction to the study of dislocations in metals and other crystalline solids. Unfortunately, because of the relatively poor penetrating power of electrons, the specimens must be in the form of very thin films and the results obtained may not be representative of the materials in bulk.

### 8.2.3 *Upper yield point*

The edge dislocation is a region of large elastic stress. At LBC in figure 62(*a*) the material is heavily compressed, at MF strongly extended. If impurity atoms are present in the lattice they will tend to concentrate at regions where they can relieve the stress. For example a *small* impurity atom will tend to take the place of the original atom B. Quite often the impurity atom forms a strong bond with the neighbouring atoms. As a result it may be more difficult to move the dislocation and detach it from the impurity. The impurity at B will thus tend to 'pin' the dislocation and a larger stress than normal will be required to make it move. Once the

† See J. W. Menter, *Proceedings of the Royal Society*, A236 (1956) p. 119, and F. P. Bowden & D. Tabor, *Friction and Lubrication of Solids* (Clarendon Press, 1964), plate III.

dislocation has moved across to position C or D it is in its normal environment and is now much easier to move. This explains the behaviour, for example, of certain mild steels where the dislocations are pinned by interstitial carbon or nitrogen. When such materials are tested in tension or compression or shear they show an 'upper yield point' (see figure 63 point B), then a region of fairly constant lower yield stress (CD) followed by a conventional work-hardening curve. If at some point F the stress is removed and the metal is 'aged' the carbon and nitrogen can diffuse to the dislocations and again pin them. The further deformation of the specimen may then show a second 'upper yield point' (G).

### 8.2.4 *Work-hardening*

Dislocations interact with one another. Continued deformation produces interlocking of dislocations and makes it more difficult for them to move. Although the details are still the subject of discussion this is the basic reason for work-hardening as a result of deformation.

### 8.2.5 *Screw dislocations*

Apart from edge dislocations there are 'screw' dislocations as shown in figure 64(*a*). If we follow a sequence of atoms, say, in the top plane starting from A and moving to BCDEF we end up at the point F which is an integral number of atomic spaces below A. This is essentially a screw movement. If the dislocation line XY moves towards CD the slipped region AXYZ increases in area

Figure 63. Schematic diagram showing upper yield stress at B and a second upper yield at G which occurs after 'ageing'.

### 8.2.6 *Crystal growth*

The presence of sites for nucleation is very important in crystal growth. On a perfectly smooth atomic plane an atom can find only one point of attachment. On a step, each atom can find two points of attachment (figure 64(*b*)). Growth will therefore occur very much more easily until the whole of the step ABCD is filled, when the surface is then plane. A screw dislocation, however, provides a self-perpetuating growth step; as further atoms become attached to the step they build up a spiral 'staircase' and so provide further steps for further condensation. This mechanism accounts for the growth spiral that is often observed on the face of crystals (see figure 64(*c*)).

## 8.3 **Vacancies, diffusion and creep**

### 8.3.1 *Vacancies*

Before leaving the ductile properties of solids we refer briefly to a few related themes which are involved in dislocation movement.

A vacancy is a place in the lattice where an atom is missing. To create a vacancy in a crystal a number of bonds have to be broken: thus a vacancy

Figure 64. (*a*) Screw dislocation, (*b*) crystal growth on an edge providing two points of attachment at E and F, (*c*) crystal growth on the edge of a screw dislocation which provides a self-perpetuating growth step.

implies a site where the atom has sufficient energy to escape. Suppose the energy required to create a vacancy is $\varepsilon_v$. If there are $n_0$ atoms in the bulk with their normal number of neighbours (i.e. ignoring the atoms in the free surface) the number $n$ possessing extra energy $\varepsilon_v$ will be

$$n = n_0 \exp(-\varepsilon_v/kT). \tag{8.12}$$

We see that vacancy-concentration increases with temperature, thus increasing the energy of the crystal. At first sight it seems strange that a stable situation exists involving a higher energy than one in which there are no vacancies. The reason is that the entropy or disorder is increased by the presence of vacancies and this compensates for the increase in energy.

It is easy to see how this arises. If initially the lattice contains no vacancies the $n_0$ atoms have $n_0$ sites to accommodate them. If the atoms are indistinguishable there is only one way in which they can be distributed, i.e. $W_1 = 1$. If $n$ vacancies are created, the $n_0$ atoms involved being ultimately located on the free surface of the specimen, we have a total of $(n_0 + n)$ sites. The number of ways in which $n_0$ indistinguishable atoms and $n$ indistinguishable vacancies can be distributed is then simply

$$W_2 = \frac{(n_0 + n)!}{n_0! n!}. \tag{8.13}$$

Thus the increase in entropy is

$$\Delta S = k \ln W_2/W_1 = k(\ln(n_0 + n)! - \ln n_0! - \ln n!). \tag{8.14}$$

By Stirling's approximation for large numbers, $N! = (N/e)^N$ or

$$\ln N! = N \ln N - N. \tag{8.15}$$

Equation (8.14) thus gives

$$\Delta S = k \left[ n_0 \ln \left( \frac{n_0 + n}{n_0} \right) + n \ln \left( \frac{n_0 + n}{n} \right) \right]. \tag{8.16}$$

Since $n \ll n_0$ (even at the melting point $n$ is usually less than $10^{-4} n_0$) we can put $(n_0 + n)/n_0 \approx 1$ and $(n_0 + n)/n \approx n_0/n$.

Equation (8.16) then gives

$$\Delta S = kn \ln \left( \frac{n_0}{n} \right). \tag{8.17}$$

The volume change is so small that the $PV$ term in the Gibb's free energy may be ignored. We may then write (see equation (2.28))

$$\Delta G = \Delta U - T\Delta S = 0 \text{ at equilibrium}$$

or

$$T\Delta S = \Delta U = n\varepsilon_v, \tag{8.18}$$

since $\varepsilon_v$ is the energy to create one vacancy. Combining equation (8.18) with equation (8.17) gives

$$kTn \ln\left(\frac{n_0}{n}\right) = n\varepsilon_v$$

or

$$\ln\left(\frac{n_0}{n}\right) = \varepsilon_v/kT. \tag{8.19}$$

This is the same as equation (8.12).

### 8.3.2 *Diffusion*

Atoms can move into a vacancy and leave their own site vacant. The vacancy has thus been displaced. This is the essential mechanism of diffusion in a solid. The energy for a neighbour to move into a vacant site will generally be less than $\varepsilon_v$, say $\varepsilon_d$. If we treat the crystal as an Einstein solid with an atomic vibrational frequency $\nu$, an atom has a chance of making

$$\nu \exp(-\varepsilon_d/kT) \tag{8.20}$$

jumps per second with this excess energy. If this is multiplied by the number of vacancies available for the atoms to jump into we obtain the number $\mathcal{N}$ of vacancy displacements per second.

$$\mathcal{N} = \nu \exp(-\varepsilon_d/kT)n_0 \exp(\varepsilon_v/kT)$$

$$= n_0\nu \exp\left(-\frac{\varepsilon_v+\varepsilon_d}{kT}\right). \tag{8.21}$$

To convert this to the diffusion coefficient $D$ we note that for a gas $D = \lambda c/3$, where $\lambda$ is the mean distance between collisions. But $c = \mathcal{N}\lambda$, where $\mathcal{N}$ is the frequency of molecular collisions. Thus for a gas $D = \frac{1}{3}\mathcal{N}\lambda^2$ and this type of relation holds for all diffusion processes although the numerical constant may vary depending on the details of the diffusion mechanism. For the crystal $\lambda = a$. Hence

$$D = \text{const} \times n_0\nu a^2 \exp\left(-\frac{\varepsilon_v+\varepsilon_d}{kT}\right)$$

$$= Cn_0\nu a^2 \exp(-q/kT), \tag{8.22}$$

where $q$ is the activation energy for self diffusion.

### 8.3.3 *Creep*

If a crystalline solid is deformed plastically under constant stress it will work-harden by the generation and interaction of dislocations until no further plastic flow occurs. If, however, the dislocations can escape from whatever obstacles impede their motion, flow at a vastly reduced rate will occur at constant stress. This is known as creep. The creep rate is usually a function of time, stress and temperature.

At temperatures above about 0.4 $T_m$ ($T_m$ = absolute melting point) the main mechanism which enables this to occur is the diffusion of atoms such that the dislocations can re-establish themselves in some slip plane where they are not blocked. In this case the creep rate, i.e. the rate of change of strain $\dot{\varepsilon}$ is constant with time. This is known as steady state or viscous creep and follows a law of the form

$$\dot{\varepsilon} = A(\text{stress})^n \exp(-q/kT). \tag{8.23}$$

In such a situation $q$ turns out to be the activation energy for self-diffusion in the solid.

### 8.4 Brittle solids

#### 8.4.1 *Brittle properties*

We now consider the behaviour of a solid which normally shows no ductility. If it is extended it stretches elastically and then snaps at some critical tensile stress $S$. If the material is homogeneous it will crack along a plane normal to the direction of the stress. If the specimen is subjected to a hydrostatic pressure $P$ the tensile stress necessary to cause brittle failure is found to be $P + S$ (figure 65). This implies that brittle failure occurs

Figure 65. Brittle failure: (*a*) in pure tension the specimen fails for a tensile stress $S$, (*b*) when a hydrostatic pressure $P$ is applied brittle failure occurs for a tensile stress of $P + S$.

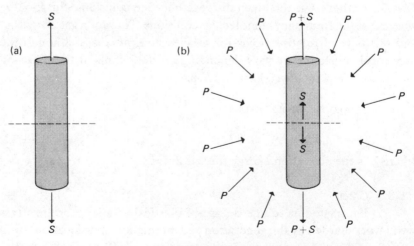

when the resulting tensile stress across some appropriate plane exceeds a critical value, in this case $S$. In its idealized form this is the stress necessary to pull one plane of atoms completely away from a neighbouring plane.

We may estimate the theoretical brittle strength in a number of ways. If the bar has unit cross-section and $\gamma$ is the surface energy, the act of snapping the specimen is to create two new unit areas of surface: the work done is $2\gamma$. If intermolecular forces are appreciable only over the distance of an atomic spacing, say $5 \times 10^{10}$ m, and the stress over this distance is assumed constant, we may write

$$\text{work done} = S \times 5 \times 10^{-10} = 2\gamma. \tag{8.24}$$

For paraffin wax, where $\gamma$ is of the order of 100 mJ m$^{-2}$, this gives a value of $S$ of about 400 MN m$^{-2}$ or 40 kg mm$^{-2}$. For rock-salt, for which $\gamma$ is about 500 mJ m$^{-2}$, the theoretical value is of order 200 kg mm$^{-2}$. This is enormously larger than the observed strength.

We may also estimate the strength from the force–separation curve (figure 66($a$)). When the force reaches the point B it is great enough to pull the atoms apart. This occurs when $\partial F/\partial x = 0$. If we express the energy $U$ in terms of $x$ we have $\partial U/\partial x = -F$. Hence failure occurs when

$$\frac{\partial^2 U}{\partial x^2} = 0. \tag{8.25}$$

We have not derived an expression for the energy of a crystal in terms of the atomic separation in one dimension. We may, however, estimate the

Figure 66. ($a$) Force–displacement curve for calculating the theoretical brittle strength, ($b$) role of a crack, length $l$ tip radius $\rho$, in facilitating brittle failure, ($c$) the formation of a crack by the 'pile-up' of dislocations.

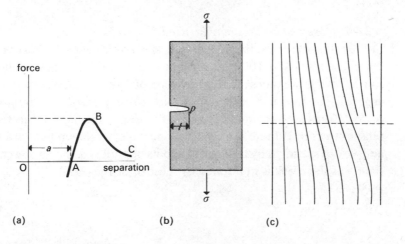

(a)    (b)    (c)

hydrostatic tension necessary to pull a crystal apart. For an ionic crystal we have (from the previous chapter)

$$U = -\text{const} \times \left(\frac{1}{x} - \frac{a^8}{9x^9}\right). \tag{8.26}$$

Ignoring the constant and differentiating,

$$\frac{\partial^2 U}{\partial x^2} = -\frac{2}{x^3} + \frac{10a^8}{x^{11}} = 0,$$

$$x = 5^{\frac{1}{8}} \times a = 1.23a. \tag{8.27}$$

This shows that the crystal can be expanded by a hydrostatic tension until when the ionic separation has increased by 23 per cent, i.e. the volume has increased by 85 per cent, it will disintegrate. If the bulk modulus $K$ remained constant for these large strains it would imply a pressure equal to 0.85 $K$, but because of the large curvature of the force separation curve in this region the stresses are appreciably smaller. Simple calculation shows that this should be reduced by a factor of about 3.5 so that the corresponding pressure is about 0.24 $K$. Using the theoretical value of $K$ from the previous chapter, this is 7000 MN m$^{-2}$ or 700 kg mm$^{-2}$. This is of the same order as that deduced from the surface energy. In practice a hydrostatic tensile experiment would be difficult if not impossible to carry out; most estimates based on direct tension give a theoretical value for the brittle strength of the order of

$$S \approx \frac{E}{10}, \tag{8.28}$$

where $E$ is Young's modulus.

### 8.4.2 *Observed brittle strength: need for cracks*

The observed brittle strength of solids is generally very variable but is usually 10 to 100 times smaller than the theoretical value. The source of this weakness is the presence of flaws or cracks in the solid, especially at the surface. These act as 'stress-raisers' or 'stress-multipliers'. Consider a rectangular strip of uniform thickness subjected to a tensile stress $\sigma$. If there is a crack in one edge, as shown in figure 66($b$), and we can specify it by its length $l$ and its tip radius $\rho$, the tensile stress at the tip of the crack is multiplied by a factor of the order

$$\left(\frac{l}{\rho}\right)^{\frac{1}{2}}. \tag{8.29}$$

Hence if the applied stress is $\sigma$ and the theoretical breaking stress is $S$, we may write for the stress to produce fracture:

$$\sigma \approx S\left(\frac{\rho}{l}\right)^{\frac{1}{2}}. \tag{8.30}$$

If the crack radius is, say, 1 nm, a crack length of 1000 nm (1 $\mu$m) will increase the tensile stress at the tip of the crack by 30-fold. Thus an applied stress only one-thirtieth of the theoretical strength will be able to start the crack growing. Once this occurs the whole section will fail.

There are three general observations which support this explanation. First, brittle strengths are usually very variable; this is because of the large variety of sizes and shapes of surface cracks. Secondly, in recent years it has proved possible to prepare fine silica rods with very perfect surfaces which can withstand a tensile stress of $E/20$ before they fail. Thirdly, some fine crystalline fibres have been prepared which have tensile strengths approaching their theoretical value, presumably because their surfaces are free of surface flaws or growth steps.

Although this explanation is generally satisfactory there is an additional failure criterion which must be mentioned. Consider a flaw which is in the form of a sharp crack between two neighbouring atomic planes; the crack-tip radius tends to zero. Is then the stress-concentration factor infinite, and does the material now possess zero strength? The answer was given by A. A. Griffith in 1921. He showed that for a crack to grow it is not sufficient for the stresses at the crack-tip to exceed the theoretical strength; in addition sufficient elastic energy must be released from the system to provide the extra surface energy that a growing crack demands. By expressing the surface energy in terms of the elastic constants, the force–separation curve, and the atomic spacing, it is possible to estimate what this involves. It turns out that an infinitely sharp crack cannot grow with the application of a vanishingly small tensile stress.[†] In order to satisfy the surface energy criterion, the smallest stress $\sigma_G$ capable of producing crack propagation, if the lattice spacing is $a$, is of order

$$\sigma_G \approx S\left(\frac{3a}{l}\right)^{\frac{1}{2}}, \tag{8.31}$$

that is to say, because of the surface energy criterion an infinitely sharp crack produces a reduction in strength equivalent to a tip radius of about three atomic spacings. For a crack of length $l$, the Griffith stress $\sigma_G$, is the

---

[†] See A. H. Cottrell, *The Mechanical Properties of Matter* (Wiley, 1964), and E. Orowan, 'Fracture and strength of solids', *Reports on Progress in Physics*, vol. 12 (1949), pp. 185–232.

smallest stress that will start the crack growing. This is the stress for a crack of tip radius $3a$ or less. For a more blunt crack ($\rho > 3a$) a larger stress is needed to start the crack moving. If the geometry of the tip is retained during propagation the failure stress falls slowly as the crack length increases. If, however, the crack grows into a sharp crack the failure stress falls to the Griffith value.

Equation (8.31) may also be expressed in terms of Young's modulus $E$ and the surface energy $\gamma$ of the solid in the following way. The work done in pulling an ideally brittle solid specimen of unit cross-section apart is equivalent to the work done in creating two new unit surfaces, i.e. $2\gamma$. As we have seen, this pulling apart occurs when the applied stress is of order $E/10$ (see equation (8.28). The separating atomic planes are pulled apart at this stage by a distance of the order $a/10$. The work done is of order $(E/10) \times (a/10) = Ea/100$, which can be equated to $2\gamma$. Hence $\gamma$ is of order $Ea/200$. Equation (8.31) then becomes

$$\sigma_G = \frac{E}{10}\left(\frac{3 \times 200\gamma}{El}\right)^{\frac{1}{2}} = \left(\frac{600}{100} \times \frac{\gamma E}{l}\right)^{\frac{1}{2}},$$

A more rigorous analysis shows that the numerical factor inside the brackets is nearer unity than 6. The Griffith stress then becomes

$$\sigma_G \approx \left(\frac{\gamma E}{l}\right)^{\frac{1}{2}}. \tag{8.32}$$

The equilibrium spreading of a crack thus provides a means of determining the surface energy of a solid. We have already described one example of this in the splitting of a mica sheet. Attempts have also been made to apply this method to metals, polymers and other solids. Generally such experiments give extremely high values of the order of $100\ \mathrm{J\ m^{-2}}$; this is because plastic deformation occurs at the tip of the crack and the plastic work involved completely swamps the much smaller true surface energy. Recently Gilman has shown that with some ionic crystals it is possible to 'freeze-out' the ductility by carrying out the experiments with the specimens immersed in liquid nitrogen. Under these conditions surface energies of the order of a few hundred $\mathrm{mJ\ m^{-2}}$ are obtained.

Edge dislocations are too small to act as effective stress-raisers from the point of view of crack propagation. However, if the free movement of edge dislocations is obstructed by some barrier so that several can pile up in a row (see figure 66($c$)), an internal crack may be formed; this may be able to initiate crack propagation. In this way a ductile solid may become brittle.

### 8.4.3 *How brittle solids may be made ductile*

Consider a brittle solid which fails in a brittle manner for a tensile stress $S$ (see figure 67). If we apply a hydrostatic pressure the tensile stress necessary for failure is $P + S$. Associated with this tensile stress is a shear stress equal to $\frac{1}{2}(P + S)$. If the critical shear stress of the material is less than this it will flow in a ductile manner before the tensile stress is large enough to produce brittle failure. This is one of the reasons why rocks below the earth can flow in a ductile way although they are normally very brittle materials. Indeed Bridgman has shown in the laboratory that under sufficiently high hydrostatic pressure even quartz can flow plastically. In this connection it is interesting to note that in indentation hardness experiments plastic indentation can often be made in relatively brittle materials (even though some cracking may also occur). This is because the large hydrostatic component of the stress field inhibits brittle failure. Furthermore the hardness values so obtained are a measure of the *plastic* properties of the brittle solid.

### 8.5 **Conclusion**

In this chapter we have discussed the strength and deformation properties of solids in terms of interatomic forces. The deformation characteristics fall into three main categories. When the stresses are

Figure 67. Stress–strain curve for a brittle solid. Brittle failure occurs for a tensile stress $S$. In the presence of a hydrostatic pressure $P$ this is increased to $P + S$. The associated shear stress may produce plastic yielding before brittle failure occurs.

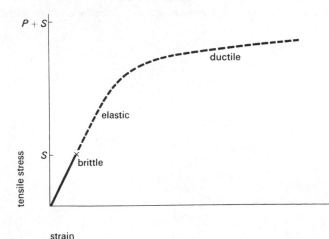

strain

below a certain level the strains are reversible and the deformation is said to be elastic. As we saw in the previous chapter elastic deformation involves the gentle distortion of the lattice and when calculations are possible there is good agreement between the observed behaviour and that derived from the interatomic forces. For larger stresses two other types of deformation occur, plastic and brittle: both are irreversible. These strength properties may also be calculated in terms of interatomic forces but observed values are generally very much smaller. We may say that solids owe their plastic and brittle strength to interatomic forces: their weakness to the presence of flaws or imperfections. The main characteristics are summarized in Table 25.

Table 25.  *Plastic and brittle failure*

| Mode | Criterion | Source of weakness | Effect of hydrostatic pressure |
| --- | --- | --- | --- |
| Plastic | Critical shear stress for planes to slide over one another | Dislocations, enable slip to occur row by row of atoms | None |
| Brittle | Tensile stress across a plane pulling planes of atoms apart | Cracks, act as stress-multipliers so that stress at crack-tip is much greater than the applied stress | Opposes applied tensile stress and so can inhibit brittle failure |

# 9

---

# *Thermal and electrical properties of solids*

In this chapter we shall discuss in terms of atomic mechanisms the specific heat capacity of solids, thermal expansion and thermal and electrical conductivity.

## 9.1 Specific heat capacity
### 9.1.1 *Definition of specific heat capacity*

The specific heat capacity is defined as the quantity of heat required to raise the temperature of unit mass of substance by 1 K. With solids the specific heat capacity at constant volume $C_V$ is a little less than that at constant pressure $C_P$. For one mole the corresponding quantities are related by the thermodynamic function

$$C_P - C_V = \beta^2 TVK, \tag{9.1}$$

where $\beta$ is the volume coefficient of expansion, $K$ the bulk modulus and $V$ the volume of one mole. In practice the specific heat capacity generally measured with solids is $C_P$, but the difference between this and $C_V$ is small and can as a first approximation be neglected.

If we assume that the specific heat capacity is due to the change in internal energy we may easily calculate it, say, for one mole of a metal. This contains $N_A$ atoms. The atoms have no translational energy, only vibrational. Each vibrational degree of freedom involves both potential and kinetic energy so that the average thermal energy is $kT$ per degree of vibrational freedom. Each atom has three independent degrees of vibration so that its average vibrational energy is $3kT$. For all $N_A$ atoms this gives a value of

$$U = N_A = 3kT = 3RT. \tag{9.2}$$

If we keep the volume constant any additional heat goes solely in increasing the vibrational energy. Hence

$$C_V = \left(\frac{\partial U}{\partial T}\right)_V = 3R, \tag{9.3}$$

where $C_V$ is the specific heat capacity per mole (atomic). Thus $C_V$ should have a value of about $25 \text{ J K}^{-1}$, a conclusion reached empirically by Dulong and Petit in 1819. If one deals with a compound, say NaCl, then one would expect each ion-mole to possess a specific heat capacity of $25 \text{ J K}^{-1}$; so that for the mole the specific heat capacity would be $50 \text{ J K}^{-1}$; for a triatomic molecule it would be $75 \text{ J K}^{-1}$. Some typical results for $C_P$ at room temperature for metals and diatomic and triatomic solids are given in Table 26. Ignoring the difference between $C_P$ and $C_V$, it is seen that the specific heat capacities per mole are all indeed of the order of $25 \text{ J K}^{-1}$ (except ice).

Table 26. *Specific heat capacities of solids*

| Substance | $C_P$ ($\text{J kg}^{-1}\text{ K}^{-1}$) | Atomic or molecular weight (kg) | Specific heat capacity $C_P \text{ mol}^{-1}$ ($\text{J K}^{-1}$) |
|---|---|---|---|
| Al | 920 | 0.027 | 25 |
| Cu | 380 | 0.063 | 24 |
| Ag | 220 | 0.108 | 24 |
| Pb | 120 | 0.207 | 25 |
| NaCl | 880 | 0.0575 | 51 (diatomic) |
| CaF$_2$ | 920 | 0.0780 | 72 (triatomic) |
| SiO$_2$ | 1180 | 0.0600 | 71 (triatomic) |
| Ice | 2100 | 0.018 | 38 (triatomic) |

### 9.1.2 *The Einstein model*

Later measurements showed that the Dulong and Petit law holds only above a certain temperature. At lower temperatures $C_P$ falls off markedly. For example with copper the specific heat capacity per mole falls off from 24 at room temperature to 21 at 170 K, to 11 at 70 K and to $1 \text{ J K}^{-1}$ at 20 K. In the limit it approaches zero at the absolute zero.

The first simple explanation of this was due to Einstein in 1906 who made use of Planck's earlier discovery of the distribution of thermal energy in an oscillating system. If each atom behaves as an independent harmonic oscillator of frequency $\nu$, its average energy (ignoring zero-point energy) is exactly the same as that given in our discussion of the vibrational energy of a diatomic molecule.

$$u = \frac{h\nu}{\exp(h\nu/kT)-1}. \tag{9.4}$$

For the $N_A$ atoms each of which has 3 independent degrees of vibrational freedom we obtain

$$U = 3N_A u. \tag{9.5}$$

For high temperatures this reduces to $3N_A kT$ in agreement with the Dulong and Petit law. For lower temperatures it diminishes and as $T$ tends to zero, it also tends to zero. If an appropriate value of $\nu$ is assumed the shape of the Einstein theoretical curve agrees reasonably well with experiment except towards $T = 0$ (see figure 68). In this region careful experiments show that $C_V$ falls of as $T^3$, whereas in Einstein's theory it falls off as $\exp(-h\nu/kT)$, i.e. it falls off more rapidly than $T^3$.

### 9.1.3 *The Debye model*

There are two main defects in the Einstein model. First it assumes that all the atoms have the same single frequency $\nu$, and secondly that the vibration of each atom is independent of its neighbour. In fact the atoms act as coupled oscillators and a whole range of frequencies is possible. Each value of $\nu$ (at a fixed temperature) contributes its own

Figure 68. Specific heat capacity of copper as a function of temperature: ● experimental values, —— theory given by Einstein assuming an atomic vibrational frequency of $\nu = 5 \times 10^{12}\,\text{s}^{-1}$. If one writes $h\nu_E/k\theta_E = 1$ this gives for the Einstein temperature, corresponding to $\nu = 5 \times 10^{12}\,\text{s}^{-1}$, a value of $\theta = 240\,\text{K}$.

average thermal energy $u$ (see equation (9.4). These should all be added to determine the total energy $U$ of the whole solid. An analysis along these lines was first developed by Born and von Karman in 1912. The main feature of such a treatment, which distinguishes it from the Einstein model, is that it reveals a whole spectrum of possible frequencies. The detailed analysis is, however, extremely complicated.

A completely different approach is that due to Debye (1912). The solid is treated as a continuum which, at first sight, seems a retrogressive step. However, its great merit is that it provides a frequency spectrum which is a close approximation to the true spectrum. Debye considers the way in which waves can travel through the solid. The most remarkable feature is that a *standing wave* of frequency $\nu$ behaves exactly like a quantum oscillator of frequency $\nu$ so that its thermal energy is again given by equation (9.4). This is surprising since the wave involves the vibration of all the atoms in the solid; yet its thermal energy is that of a *single* quantum oscillator. At first sight it would seem that this could not possibly give the same total energy. The difficulty, however, disappears when it is realized that the total vibration of each atom is the composite result of all the possible waves that can travel through the solid.

Suppose the solid is a crystal in the form of a cube of side $L$. To establish standing waves the free surface of the specimen must be either a node or an antinode and it may be shown that either condition leads to the same result. We shall treat the free surface as an antinode. Then the largest

Figure 69. Characteristic frequencies of standing waves in an elastic continuum of length $L$. We assume that the free ends are antinodes and that the velocity $C$ of the waves is independent of frequency.

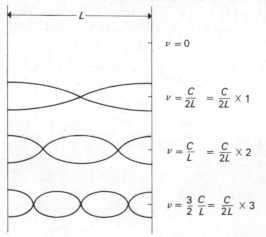

$\nu = 0$

$\nu = \dfrac{C}{2L} = \dfrac{C}{2L} \times 1$

$\nu = \dfrac{C}{L} = \dfrac{C}{2L} \times 2$

$\nu = \dfrac{3}{2}\dfrac{C}{L} = \dfrac{C}{2L} \times 3$

standing wave possible for a wave travelling normal to the faces of the cube has a wavelength $\lambda = 2L$ (see figure 69). If the velocity of the wave is $C$, the frequency of the wave is $\nu = C/\lambda = C/2L$. The next possible standing wave has a wavelength $\lambda = L$ and its frequency is $\nu = C/L$. The next has a wavelength $\lambda = \frac{2}{3}L$ and its frequency is $\nu = \frac{3}{2} C/L$. Thus the possible frequencies increase in the order:

$$\frac{C}{2L}; \quad \frac{C}{2L} \times 2; \quad \frac{C}{2L} \times 3; \quad \frac{C}{2L} \times 4, \quad \text{etc., or } \nu_1, 3\nu_1, 4\nu_1.$$

$$(9.6)$$

The corresponding thermal energies are

$$\frac{h\nu_1}{\exp(h\nu_1/kT)-1}, \quad \frac{2h\nu_1}{\exp(2h\nu_1/kT)-1}, \quad \frac{3h\nu_1}{\exp(3h\nu_1/kT)-1}, \quad \text{etc.,}$$

$$(9.7)$$

and all that is now needed is to sum them. Very soon the possible frequencies become such large multiples of $\nu_1$ that they become virtually continuous and the sum can be replaced by an integral. One needs to know the number of vibrations that are possible between $\nu$ and $\nu + d\nu$. This is not difficult for a continuum. A simple approach is as follows.

For the waves discussed above

$$\nu_n = \frac{C}{2L} n,$$

where $n$ is an integer. If we considered the more general case of a wave travelling in some arbitrary direction we would obtain the same result but $n$ would have components $n_1, n_2, n_3$ (each of them integers) in the $x, y, z$ directions.

Then

$$\nu = \frac{C}{2L} (n_1^2 + n_2^2 + n_3^2)^{\frac{1}{2}}. \tag{9.8}$$

We plot $n_1 C/2L, n_2 C/2L, n_3 C/2L$ on $x, y, z$ coordinates (see figure 70). Then the distance from the origin to any point is equal to $\nu$. Since $n_1, n_2, n_3$ can only change by one unit at a time each possible point for $\nu$ occupies a volume of $(C/2L)^3$. The number of points available for frequencies between $\nu$ and $\nu + d\nu$ is the volume of the spherical shell contained between radii $\nu, \nu + d\nu$ divided by the unit volume $(C/2L)^3$. Since $\nu$ can have only positive values we can only use one-eighth of the spherical shell. The resulting number is therefore

$$d\mathcal{N} = (\tfrac{1}{8} \times 4\pi \nu^2 \, d\nu)\left(\frac{2L}{C}\right)^3$$

$$= \frac{4\pi L^3 \nu^2 \, d\nu}{C^3}. \tag{9.9}$$

We allow for two transverse vibrations and one longitudinal vibration. These have different wave velocities but for simplicity we choose a suitable average velocity $C$. This triples the value of $d\mathcal{N}$. Then the total thermal energy is simply

$$U = \int_0^{\nu_m} \left[\frac{h\nu}{\exp(h\nu/kT) - 1}\right] 3 \, d\mathcal{N}. \tag{9.10}$$

We have ignored zero-point energy since it is independent of $T$ and when we differentiate $U$ to find $C_V$ it will disappear.

The difficulty about equation (9.10) is that we have no way, so far, of specifying the upper limit $\nu_m$ of the integral. It is at this point that Debye ties his continuum model to a particulate model. If the crystal contains $N$ atoms, Debye postulates that the total number of vibrations possible must be equal to $3N$ so that, in the limit, at high temperatures where each $u$ has

Figure 70. (*a*) Construction showing that the number of waves d$N$ possessing frequency between $\nu$ and $\nu + d\nu$ is proportional to the volume in the octet of the spherical shell lying between radii $\nu$ and $\nu + d\nu$. (*b*) Two-dimensional figure showing that each point corresponding to a particular value of $\nu$ occupies a cube of side $C/2L$.

a value of $kT$ the total thermal energy is $3NkT$. From equation (9.9) this means that

$$\int_0^N 3 \, d\mathcal{N} = \int_0^{\nu_m} 12\pi \frac{L^3}{C^3} \nu^2 \, d\nu = 3N. \tag{9.11}$$

This gives

$$\nu_m^3 = \frac{3}{4\pi} \frac{C^3}{L^3} N, \tag{9.12}$$

where $\nu_m$ is the maximum frequency on the Debye model. Before we use this we may form some estimate of its magnitude. If the solid has a simple cubic structure with a distance $a$ between each atom, $Na^3$ will be the volume of the solid, i.e. $L^3$. Then

$$\nu_m = \left(\frac{3}{4\pi}\right)^{\frac{1}{3}} \frac{C}{a} \approx \frac{C}{a}. \tag{9.13}$$

This corresponds to a minimum wavelength

$$\lambda_{\min} = \frac{C}{\nu_m} \approx a. \tag{9.14}$$

Thus the limiting frequency on the Debye model corresponds to very short waves, the wavelength being comparable with the atomic spacing. This is reasonable since once we recognize the particulate nature of the solid, a wavelength less than the atomic spacing ceases to be meaningful.

Figure 71. Specific heat capacity of various solids plotted as a function of $kT/h\nu_m$, where $\nu_m$ is the characteristic Debye frequency. Writing $h\nu_m/k = \theta_D$, the quantity $\theta_D$ is known as the Debye temperature. $\bigcirc$, $\bullet$, $\blacksquare$, $\times$, experimental points for Cu ($\theta_D = 315$ K), Ag ($\theta_D = 215$ K), Pb ($\theta_D = 88$ K), C ($\theta_D = 1860$ K); —— theory given by Debye.

This result also corresponds rather closely to the very simple elastic model discussed in the previous chapter where we found that if the atoms have a natural (uncoupled) frequency $\nu$ the velocity of a wave is of order $2\pi\nu a$ compared with $\nu_m a$ given in equation (9.13). Inserting equation (9.12) in (9.9),

$$3 \, d\mathcal{N} = \frac{9N}{\nu_m^3} \nu^2 \, d\nu. \tag{9.15}$$

If this is substituted in equation (9.10) we have an explicit expression for $U$ which can be evaluated. By differentiating, the specific heat capacity $C_V$ may be found.

The Debye theory predicts that the heat capacity of a solid depends only on the characteristic frequency $\nu_m$. This implies that if $C_V$ for various solids is plotted against $kT/h\nu_m$ they should all lie on a single curve. This is found to be very nearly true (see figure 71). The quantity $h\nu_m/k$ is known as the Debye temperature $\theta_D$. It is interesting to compare $\theta_D$ as calculated from equation (9.12) with the value of $\theta_D$ that gives the best fit of the Debye equation with the observed specific heat capacity measurements (Table 27). Although the theory is only really valid for isotropic solids containing a single type of atom it is seen that it holds extremely well for more complicated materials. It even holds approximately for ice, which is rather surprising. (An average wave velocity of 220 m s$^{-1}$ for ice has been used to calculate $\theta_D$). The specific heat capacity of ice at 200 K is about 1400 J kg$^{-1}$ K$^{-1}$ or 25 J mol$^{-1}$ K$^{-1}$. It thus behaves as a monatomic solid, practically the whole of the thermal energy being due to oscillations of the molecule as a whole. Between 200 K and 273 K (the melting point) there

Table 27. *The Debye temperature*

| Substance | $\theta_D$ | |
| --- | --- | --- |
| | From specific heat capacity | From equation (9.12) |
| Pb | 88 | 75 |
| Cd | 168 | 174 |
| Ag | 215 | 220 |
| Cu | 315 | 341 |
| NaCl | 281 | 305 |
| CaF$_2$ | 474 | 510 |
| Ice | 220 | 200 |

is some increase in specific heat capacity to a value of about $40 \text{ J mol}^{-1} \text{ K}^{-1}$. Presumably in this temperature range rotational degrees of freedom of the whole molecule contribute to the thermal energy of the ice crystal. As we shall see in Chapter 12 this rotational freedom accounts for the rather large low-frequency dielectric constant of ice well below the melting point (see figure 108).

Although the Einstein model is in many ways a simpler and more direct one than the Debye model there are two features in the Debye treatment that make it preferable. First $\nu_m$ may be derived directly from bulk properties so that there are no assumed constants in the final calculation of $U$. By contrast the Einstein frequency (which turns out to be comparable with Debye's $\nu_m$, in fact for many solids, $\nu_E \approx 0.75 \nu_m$ or $\theta_E \approx 0.75 \theta_D$) must be deduced empirically from the shape of the $C_V - T$ curve. Secondly, at low temperatures the Debye relation for the specific heat reduces to

$$C_V = \frac{12\pi^4 R}{5\theta_D^3} \times T^3, \qquad (9.16)$$

so that it satisfactorily explains the observed $T^3$ dependence in this temperature range.

The main defect of the Debye treatment is that it considers waves of all frequencies to travel at the same speed. This is not true of coupled oscillators. For higher frequencies there is an appreciable drop in wave velocity. Fortunately the waves in this range generally contribute only a small part to the total vibrational energy.

## 9.2 Thermal expansion: Gruneisen's law

We may now derive a simple relation between thermal expansion and specific heat. It is based on the existence of an asymmetrical potential energy curve for the whole crystal in terms of the separation between atoms. For simplicity we consider here only the potential energy $u$ between one atom and its neighbour and ignore the problem of coupled interactions.

The potential energy curve is shown in figure 72 $(a)$; we transpose it to axes as shown in figure 72$(b)$ and express $u$ in terms of the displacement $x$ of the atoms from their equilibrium separation $a$. We assume that at 0 K the energy is zero (ignoring zero-point energy) and that the addition of thermal energy increases $u$ according to a power law:

$$u = Ax + Bx^2 + Cx^3 \ldots . \qquad (9.17)$$

Since O is the origin where $\partial u/\partial x = 0$ we obtain $A = 0$. Then

$$u = Bx^2 + Cx^3 + \dots . \tag{9.18}$$

If we used only $u = Bx^2$ the curve would be symmetrical about the $u$ axis and the restoring force for small displacements, $-\partial u/\partial x = -2Bx$, would be proportional to $x$, so that the atoms would oscillate with simple harmonic motion. We now assume that the cubic term is sufficient to match the real asymmetric potential energy curve.

$$u = Bx^2 + Cx^3. \tag{9.19}$$

Suppose we add energy $u$ to the system by heating the solid. We see from figure 72(b) that there are two equilibrium positions P and Q between which $x$ oscillates. In the first edition of this book we assumed that the mean position is the mid-point between P and Q. It occurs (see figures 72(b) and 73) at a displacement

$$e = -Cx^2/2B. \tag{9.20}$$

A more elegant and, physically, more attractive approach is the following due to Dr J. R. Waldram. From equation (9.19) the force between the two atoms is

$$F = -\frac{\partial u}{\partial x} = -2Bx - 3Cx^2. \tag{9.21}$$

Figure 72. (*a*) Potential energy $u$ of an atom as a function of its separation $x$ from its neighbours, (*b*) same curve transposed to different axes so that the minimum has coordinates $(0, 0)$.

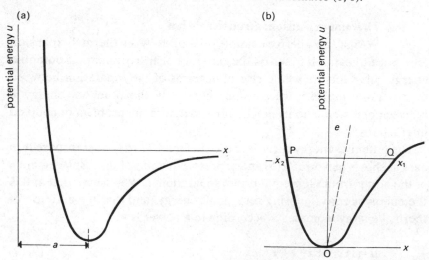

We now observe that since there is no external force acting, the time average of $F$ must be zero. Replacing $x$ and $x^2$ by $\bar{x}$ and $\overline{x^2}$ (the time average of these quantities) we thus have

$$-2B\bar{x} - 3C\overline{x^2} = 0$$

or

$$\bar{x} = \frac{-3C}{2B}\overline{x^2}. \tag{9.22}$$

The average displacement is three times as great as in the earlier derivation. This is because at Q the slope of the potential energy curve is less than at P, the restoring force is less and the rate of movement of the atom is less: it spends more time near Q than near P.

We now consider the energy as $x$ oscillates between P and Q. At P and Q all the added energy is potential ($u$): there is no kinetic energy. At position O there is no additional potential energy, all the added energy is kinetic. For any system undergoing (approximate) simple harmonic motion the average kinetic energy = average potential energy = $\frac{1}{2}u$. The potential energy is given approximately by the first term of equation (9.19), the second term representing only that small part of the energy which accounts for the skew-shaped character of the potential energy

Figure 73. ($a$) Force–separation curve for the situation described in figure 72($b$); the expansion $e$ is equivalent to an increase d$x$ in the mean spacing.

curve. The time average potential energy may thus be written as

$$\overline{Bx^2} = \tfrac{1}{2}u. \tag{9.23}$$

For a mole ($N_A$ atoms) and for a solid in which the coordination number is $n$ the increase $U$ in lattice energy will, from equation (9.23), be

$$\tfrac{1}{2}U = \overline{Bx^2}(\tfrac{1}{2}N_A n), \tag{9.24}$$

the half being introduced in the last term to avoid counting every bond twice. Then from equation (9.22) we have

$$\bar{x} = -\frac{3C}{2B^2}\frac{U}{Nn}. \tag{9.25}$$

Without any further elaboration we can see that the coefficient of linear expansion $\alpha$ which is equal to $\mathrm{d}(\bar{x}/a)/\mathrm{d}T$ will be proportional to $\mathrm{d}U/\mathrm{d}T$, i.e. to the specific heat capacity $C_V$ of the solid. Thus

$$\frac{\alpha}{C_V} = \text{constant for a given material.} \tag{9.26}$$

This relation is due to Gruneisen and we shall now derive the value of the constant more quantitatively.

We shall find it convenient to introduce the bulk modulus $K$. We assume the solid to be isotropic. Let $V$ be the molar volume. Suppose as a result of applying a hydrostatic pressure $p$ on the solid we produce a volume change $v$ corresponding to a mean change of $x$ in the separation between each atom and its neighbour ($v$ is negative).

$$K = \frac{p}{-v/V} = \frac{pV}{-v}. \tag{9.27}$$

The work done on the solid in increasing $p$ from 0 to its final value is the gain in lattice energy $U$ associated with changing the atomic spacing. The work done is

$$U = \tfrac{1}{2}p(-v) = \frac{1}{2}\frac{K}{V}(v)^2 = \frac{1}{2}KV\left(\frac{v}{V}\right)^2, \tag{9.28}$$

and since $U = \tfrac{1}{2}Nnu$ ($u$ is the energy per single bond) and $v/V = 3x/a$ we have

$$\tfrac{1}{2}Nnu = \frac{1}{2}KV\left(9\frac{x^2}{a^2}\right). \tag{9.29}$$

From equation (9.19) we see that $u \approx Bx^2$. Equation (9.29) becomes

$$BNn = 9KV/a^2. \tag{9.30}$$

We can substitute this value of $BNn$ into the denominator of equation (9.25)

$$\bar{x} = -\frac{3C}{2B}\frac{Ua^2}{9KV}.$$

The fractional volume change $v/V$ can be written

$$\frac{v}{V} = \frac{3\bar{x}}{a} = -\left(\frac{Ca}{2B}\right)\frac{1}{K}\frac{U}{V}. \tag{9.31}$$

The temperature variation of this is the volume coefficient of thermal expansion $\beta$.

$$\beta = -\left(\frac{Ca}{2B}\right)\frac{1}{K}\frac{C_V}{V} = \gamma\frac{1}{K}\frac{C_V}{V}, \tag{9.32}$$

where $C_V$ is the constant-volume molar heat capacity. The quantity $-(Ca/2B)$ is a pure number which depends only on the law of force between the atoms and is called the Gruneisen constant $\gamma$.

We may now take the final step. From equation (9.19) we see that

$$\frac{\partial^2 u}{\partial x^2} = 2B + 6Cx$$

so that

$$\left[\frac{\partial^2 u}{\partial x^2}\right]_{x=0} = \ddot{u}_0 = 2B.$$

Again

$$\frac{\partial^3 u}{\partial x^3} = \dddot{u} = 6C.$$

Consequently

$$\gamma = -\frac{Ca}{2B} = \frac{\dddot{u}}{\ddot{u}_0}\frac{a}{6}. \tag{9.33}$$

If we write out the full potential energy equation for an ionic solid as

$$u = -k\left[\frac{1}{a+x} - \frac{a^8}{9(a+x)^9}\right], \tag{9.34}$$

$\ddot{u}$ at $x = 0$ is $-k(-8/a^3)$, $\dddot{u}$ is $-k(104/a^4)$. Hence $\gamma = 13/6 \approx 2.2$.

For inert gases, assuming a 6–12 potential, the value predicted from equation (9.33) is $\gamma = 3.5$. The observed values for most solids appear to lie between about 1.5 and 3.

As an example we take an ionic solid such as rock-salt. For this $\gamma = 13/6$, $K = 2.8 \times 10^{10} \, \text{N m}^{-2}$, $V = 4 \times 10^{-5} \, \text{m}^3 \, \text{mol}^{-1}$, $C_P \approx C_V = 50 \, \text{J K}^{-1} \, \text{mol}^{-1}$. Substituting in equation (9.32) we obtain a value

$$\beta \approx 10^{-4} \, K^{-1}.$$

This is the right order of magnitude.

As has been pointed out at the end of Chapter 7 the above treatment has one basic defect: it treats the solid as though its behaviour were exactly analogous to that of a pair of atoms. By contrast, for an assembly of a large number of atoms, as in a solid crystal, an entropy factor arises. This occurs in the following way. Because of the shape of the force–separation curve an expansion involves a decrease in the force constant between individual atoms. This leads to a decrease in the natural frequency $\nu$ of the atoms and this in turn implies a smaller energy quantum $h\nu$. Since at constant temperature, in the classical region, the total vibrational energy is constant ($u = 3kT$ per atom) this means that the number of vibrational quanta in the material is increased by the expansion. The more quanta there are, the more ways there are of distributing them randomly, i.e. there is an increase in entropy. Consequently the entropy term favours expansion. On the other hand a change in atomic spacing necessitates an increase in the stored elastic energy and is therefore 'opposed' by the elastic bulk modulus of the material. A balance is struck at a certain value of the coefficient of thermal expansion $\beta$. It may be derived from the Helmholtz free energy which contains two terms, one being the stored elastic energy and the other the entropy term. When this analysis is carried out it is found that, to a very close approximation,

$$\beta = \gamma \frac{1}{K} \frac{C_V}{V},$$

as in equation (9.32), where $\gamma = -\text{d}(\ln \nu)/\text{d}(\ln V)$. The natural frequency is proportional to (force constant)$^{\frac{1}{2}}$, i.e. to $(\partial^2 u/\partial x^2)^{\frac{1}{2}}$. The volume $V$ is proportional to $a^3$ so that $\text{d}(\ln V)$ is equal to $3\text{d}(\ln a)$. Using equation (9.19) for $u$ we find that

$$\gamma = -\frac{\text{d}(\ln \nu)}{\text{d}(\ln V)} = -\frac{Ca}{2B}.$$

This is identical with the result obtained above.

Finally we may note that even if the atomic vibrations were perfectly harmonic, i.e. if the potential energy curve were perfectly symmetrical, a notional increase in the atomic spacing would reduce the restoring force on a displaced atom and the vibrational frequency would be correspondingly reduced. This would result in a finite coefficient of thermal expansion (Guggenhcim, 1965†). However, the thermal expansion due to this factor would be very small compared with that arising from the asymmetric potential energy curve considered above.

### 9.3 Thermal conductivity

#### 9.3.1 *Heat transfer*

There are three main forms of heat transfer.

(*a*) Radiation; this involves the emission of electromagnetic waves. All materials can radiate but radiation is generally most marked with solids.

(*b*) Convection; this involves the streaming of matter. With forced convection the streaming is imposed by some external driving mechanism; with natural convection it occurs as a result of density differences. Convection is observed only in liquids and gases.

(*c*) Conduction; this involves the transfer of thermal energy by some form of collision. It occurs in gases, liquids and solids. With poorly conducting solids the energy transfer is provided by lattice vibrations; with metals this is greatly augmented by electron collisions. In this chapter we shall deal solely with conduction in solids.

The rate at which heat flows through any element of a solid is proportional to the cross-section area d$A$ of the element and the temperature gradient measured normal to d$A$. We may make this proportionality into an equality by inserting a constant $K$ which we call the thermal conductivity. Then

$$\frac{dQ}{dt} = -K \, dA \times \frac{dT}{dx}. \tag{9.35}$$

The negative sign shows that a positive flow of heat takes place from a hotter to a cooler region, i.e. where d$T$/d$x$ is negative. $K$ has the dimensions $MLT^{-3}\theta^{-1}$. It varies from the best thermal insulators to the best conductors by a factor of only about 1000. By contrast, electrical conductivities can vary by a factor of $10^{30}$.

#### 9.3.2 *Heat flow down a uniform bar*

As a simple example we consider the heat flow down a uniform bar under two extreme conditions. When it is:

(*a*) thermally insulated. In the steady state if no heat escapes from the surface of the bar the quantity of heat crossing every section per second

† E. A. Guggenheim, *Boltzmann's Distribution Law* (North-Holland, 1965).

must be constant. The temperature gradient must therefore be constant (see equation (9.35)). Knowing the temperature of the hot and cold ends, the cross-section of the bar and the rate of heat conduction, $K$ may be determined directly. This is a convenient method of measuring $K$ for good conductors. It may also be used with poor conductors if the bar is replaced by a thin disc of the material.

($b$) exposed to a cooler constant temperature environment. Consider a very long bar of uniform cross-sectional area $A$ held at a temperature $T_0$ at one end (figure 74). Let the environment be at a temperature $T_e$. Consider a small element of length $dx$ at some point along the bar where its temperature is $T$. Express these temperatures as temperature excesses over the environmental temperature. The hot end is at $\theta_0$, the element at $\theta$ and the environment at zero. In the steady state heat enters at $A$, some

Figure 74. ($a$) Steady-state flow of heat along an exposed uniform bar. The heat flow in at A, less the heat flow out at B is equal to the heat lost by radiation and convection from the surface between A and B. ($b$) Heat flow through a thermally insulated element: the heat flow in at A, less the heat flow out at B is equal to the heat accumulated in the element.

is lost by radiation and convection from the surface of the element, and the remainder passes through the element at $B$. If the circumference of the bar is $c$ the surface area of the element is $c\,dx$. If the heat lost is proportional to the surface area, the emissivity $\varepsilon$ of the surface, and the temperature excess over the surroundings, the heat lost per second is $c\,dx\varepsilon\theta$.

Heat flow per second in at $A = -KA\dfrac{d\theta}{dx}$.

Heat flow per second out at $B = -KA\dfrac{d}{dx}(\theta + d\theta) = -KA\left(\dfrac{d\theta}{dx} + \dfrac{d^2\theta}{dx^2}dx\right)$.

Heat flow per second at A exceeds outflows at B by

$$KA\dfrac{d^2\theta}{dx^2}dx = \text{heat lost from surface of element, so that}$$

$$KA\dfrac{d^2\theta}{dx^2} = c\varepsilon\theta. \tag{9.36}$$

The solution is of the form

$$\theta = B\,e^{wx} + C\,e^{-wx},$$

where $w^2 = c\varepsilon/KA$. At great distances from the hot end the temperature approaches the ambient, i.e. $\theta = 0$. Hence $B = 0$.

$$\theta = C\,e^{-wx}. \tag{9.37}$$

Since at $x = 0$, $\theta = \theta_0$ we have finally

$$\theta = \theta_0 \exp\left[-\left(\dfrac{c\varepsilon}{KA}\right)^{\frac{1}{2}}x\right]. \tag{9.38}$$

The temperature falls off exponentially with distance from the hot end.

### 9.3.3 *Thermal diffusivity*

Consider a body which is insulated so that no heat is lost to the surroundings. We heat one part of it and consider how the heat flows before steady-state conditions are reached. Clearly the body has to be warmed up (i.e. it must absorb heat) as the heat flows through it.

Consider a small element of cross-section $dA$ at a point A (ordinate $x$) where the temperature is $T$. At the neighbouring point B (ordinate $x + dx$) the temperature is $T + dT$ (see figure 74(b)).

Heat flow per second in at $A = -K\,dA\dfrac{dT}{dx}$.

Heat flow per second out at $B = -K\,dA\dfrac{d}{dx}(T + dT)$.

The heat accumulated in the element per second $= K\,dA\,(\partial^2 T/\partial x^2)\,dx$. If the element has density $\rho$, specific heat capacity $C$, its heat capacity is $\rho\,dA\,dxC$. If the temperature rise per second of the element is $dT/dt$ we have

$$\rho C\,dA\,dx\,\frac{dT}{dt} = K\,dA\,\frac{\partial^2 T}{\partial x^2}\,dx, \tag{9.39}$$

$$\frac{dT}{dt} = \frac{K}{\rho C}\,\frac{\partial^2 T}{\partial x^2}.$$

The quantity $K/\rho C$ is known as the thermal diffusivity or *temperature conductivity*. It is of great importance in non-steady-state thermal problems.

### 9.3.4 *Theory of thermal conductivity*

The thermal conductivity of solids can be explained in terms of some type of collision process by means of which thermal energy is transported from the hotter to the colder regions. It is natural therefore to express it in a manner resembling the treatment for gases. For a gas where the transfer is due to molecular collisions we have

$$K = \tfrac{1}{3}nm\bar{c}\lambda c_V, \tag{9.40}$$

where $c_V$ is the specific heat capacity per unit mass. We note that the quantity $nmc_V$ is the specific heat capacity per unit volume $s$. Then

$$K = \tfrac{1}{3}s\bar{c}\lambda. \tag{9.41}$$

9.3.4.1 *Poor conductors.* In poor conductors the thermal energy is transferred by lattice vibrations. These travel through the specimen as waves which are scattered by the lattice as they progress through it. The wave is attenuated corresponding to the drop in temperature through the specimen. These thermal waves thus behave like particles possessing energy and momentum: they are known as *phonons*.

We may treat the waves as energy packets moving with the velocity $c$ of a sound wave, and travelling a distance $\lambda$ before they collide and communicate their thermal energy to the lattice. By exact analogy with gases we may write

$$K = \tfrac{1}{3}(\text{specific heat capacity per unit volume})$$
$$\times (\text{velocity of sound wave})\lambda.$$

This relation is found to be quite satisfactory for most poor conductors, the value of $\lambda$ being of the order of 2 nm.

9.3.4.2 *Good thermal conductors: metals.* Good thermal conductors are generally good electrical conductors – their thermal conductivity is

100–1000 times greater than that of insulators. This suggests that the major part of the thermal conductivity, like that of the electrical conductivity, arises from movement of electrons.

We make the following assumptions about the metallic state:

(*a*) The metal consists of fixed positive ions in a sea of electrons: in general there will be one or two electrons per ion.

(*b*) The electrons behave as a perfect gas transporting thermal energy as in gas conduction.

(*c*) Each electron has thermal energy of translation equal to $\frac{3}{2}kT$, i.e. its thermal capacity is $\frac{3}{2}k$.

(*d*) Each electron travels a distance $\lambda$ before colliding with a positive ion and giving up all its thermal energy.

We rewrite equation (9.40) in the form

$$K = \tfrac{1}{3}n\bar{c}\lambda(mc_V),\tag{9.42}$$

and note that $mc_V$ is the thermal capacity of a single travelling particle. Replacing this by $\frac{3}{2}k$ for our electron gas, we obtain

$$K = \tfrac{1}{2}n\bar{c}\lambda k,\tag{9.43}$$

where $\bar{c}$ is the velocity of an electron assuming it to behave as a gas molecule. At room temperature its value is of order $10^5\,\mathrm{m\,s^{-1}}$. If we consider a monovalent metal such as sodium, for which there is a single electron per ion, we have $n \approx 4 \times 10^{28}\,m^{-3}$, $k = 1.4 \times 10^{-23}\,\mathrm{J\,K^{-1}}$. The observed conductivity at room temperature is about $130\,\mathrm{J\,m^{-1}\,K^{-1}}$. This gives $\lambda \approx 4\,\mathrm{nm}$ or $40\,\text{Å}$. In general, equation (9.43) gives a good agreement with observation on the assumption that $\lambda$ is of the order of a few atomic spacings. Although this seems satisfactory there is a major defect in the theory which we shall discuss later.

## 9.4 Electrical conductivity of metals

### 9.4.1 *Drude model*

The general picture that we adopt here, which is due to Drude (1912), is similar to that used in the preceding section. It is assumed that the electrons move like a perfect gas in all directions through the lattice and that their thermal motion is terminated by collisions with positive ions. Under normal conditions there is no net transfer of charge in any direction. When, however, a potential is applied across the metal the electric field imposes a drift velocity on the electrons and it is this which is responsible for the observed electrical conductance. The drift velocity is very small compared with the mean electron-gas velocity $\bar{c}$.

Consider a bar of uniform cross-section $A$, length $L$, across which a potential $V$ is applied (figure 75). The electric field on the electrons is $X = V/L$. The force on an electron is $eX$ and its acceleration is $\alpha = eX/m$. If $\lambda$ is the mean free path of the electron the time between collisions is $t = \lambda/\bar{c}$. The drift velocity acquired during this period is $\alpha t$ so that the average drift velocity $u$ between collisions is $\frac{1}{2}\alpha t$:

$$u = \tfrac{1}{2}\alpha t = \frac{1}{2}\frac{eX}{m}\frac{\lambda}{\bar{c}}. \tag{9.44}$$

The model still assumes that the electron gives up the velocity gained from the field by colliding with a positive ion. If $n$ is the number of free electrons per m$^3$ the number travelling per second along the bar due to drift is the number in a volume $Au$, i.e. $Aun$. Each electron carries a charge $e$ so that the charge carried per second, i.e. the current $i$, is

$$i = eAun$$

$$= \frac{1}{2}\frac{AX\lambda e^2 n}{m\bar{c}}, \tag{9.45}$$

or

$$i = \frac{AV}{L} \times \frac{\lambda e^2 n}{2m\bar{c}}. \tag{9.46}$$

*If* $\rho$ is the specific resistivity and $\kappa$ the electrical conductivity we know that

$$i = \frac{AV}{L} \times \frac{1}{\rho} \quad \text{or} \quad \frac{AV}{L} \times \kappa. \tag{9.47}$$

Figure 75. Electrical conductivity model for metallic conductors.

(a)                     (b)

Comparison with (9.46) shows that†

$$\kappa = \frac{\lambda e^2 n}{2m\bar{c}} = \frac{\lambda e^2 n\bar{c}}{2m\bar{c}^2}.$$ (9.48)

If we ignore the difference between $\bar{c}^2$ and $\overline{c^2}$ we may, for the electron gas, write

$$\tfrac{1}{2}m\bar{c}^2 = \tfrac{3}{2}kT.$$ (9.49)

Hence, from equation (9.48)

$$\kappa = \frac{\lambda e^2 n\bar{c}}{6kT}.$$ (9.50)

For a monovalent metal assuming $\lambda = 0.3$ nm, $n \approx 4 \times 10^{28} \, \text{m}^{-3}$, $e = 1.6 \times 10^{-19} \, \text{C}$,

$$\bar{c} = 10^5 \, \text{m s}^{-1}, \, k = 1.4 \times 10^{-23} \, \text{J K}^{-1}, \, T = 300 \, \text{K},$$

$$\kappa \approx 10^6 \, \text{ohm}^{-1} \, \text{m}^{-1}.$$ (9.51)

This is in reasonable agreement with the resistivity values of most metals, $\rho \approx (2 \text{ to } 20) \times 10^{-8}$ ohm m, or conductivity $= 5 \times 10^6$ to $5 \times 10^7 \, \text{ohm}^{-1} \, \text{m}^{-1}$.

### 9.4.2 *Wiedemann–Franz relation*

We may combine equations (9.43) and (9.50) to find the ratio of thermal to electrical conductivity. We have

$$\frac{K}{\kappa} = 3\left(\frac{k}{e}\right)^2 T.$$ (9.52)

Historically, Wiedemann and Franz observed that $K/\kappa$ is the same for all metals at the same temperature. This was extended by Lorenz who observed that the ratio was proportional to the absolute temperature. The above analysis contains both 'laws'. The theoretical ratio is indeed close to the observed values over an appreciable temperature range for a large number of metals.

### 9.4.3 *Specific heat capacity of the free electron*

In the Drude model discussed above the electron gas is assumed to possess thermal energy. Like any monatomic gas it should therefore contribute a specific heat capacity of $\tfrac{3}{2}R$. The specific heat capacity of a

---

† There is a subtle distinction between the time average ($t = \lambda/c$) and the distance average of the mean free path $\lambda$. This distinction, if followed through, leads to a value of $\kappa$ twice that given in equation (9.48). A similar factor applies to the thermal conductivity $K$ so that the ratio (see equation (9.52)) is unchanged.

monovalent metal should therefore consist of $3R$ from its vibrational (lattice) energy and $\frac{3}{2}R$ from its free electrons, i.e. the specific heat capacity should have a value of about 38 J mol$^-$ K$^{-1}$. The observed value in fact does not exceed 25 J mol$^{-1}$ K$^{-1}$.

This is a basic defect of the electron-gas model and the explanation can only be given in terms of the energy states of the electrons in a metal. The spacing of these states is independent of the size of the specimen: it depends only on the number of electrons per unit volume. Considering the energy distribution of electrons in unit volume we observe that no energy state can contain more than two electrons (of opposite spins), so that the electrons fill the available states from the lowest available energy level until they reach the topmost or Fermi level, $E_F$ (see figure 76). This is the situation at absolute zero. If the metal is heated, only those electrons which can be moved into a higher vacant state above $E_F$ can acquire further energy. In effect this means that only the electrons at the top of the Fermi level can acquire energy. Analysis shows that if there are $n$ free electrons per m$^3$ the number of thermally excited electrons at temperature $T$ is

$$\Delta n \approx n \times \frac{kT}{E_F}. \tag{9.53}$$

For a typical metal $E_F \approx 2$ eV and at room temperature ($T \approx 300$ K) $kT$ is about $4 \times 10^{-21}$ J $\approx 0.025$ eV. Thus $\Delta n$ is only about 1 per cent of $n$. If the

Figure 76. Electronic energy levels in a metal shown as $dn/dE$ plotted against the energy $E$ so that the shaded area represents the number of electrons per m$^3$ with energy between $E$ and $E + dE$. (*a*) At absolute zero all the energy states are filled up to a level $E_F$ known as the Fermi level. (*b*) At a higher temperature $T$ a small number of electrons at the top of the Fermi level are able to acquire additional energy of order $kT$.

(a)  (b)

thermal energy acquired by these electrons is of order $kT$ we see that the specific heat capacity is only about 1 per cent of the value quoted above. It is for this reason that the electronic contribution to the specific heat capacity is so small.

We may take this a stage further. Since the specific heat capacity of a single electron is of the order $k$, we see from equation (9.53) that the specific heat capacity per m$^3$ is of order

$$k \, \Delta n = \frac{nk^2 T}{E_F}, \tag{9.54}$$

so we may regard the *average* specific heat capacity per electron as

$$c_e = \frac{k^2 T}{E_F}. \tag{9.55}$$

The expression from equation (9.42) for the thermal conductivity may be written

$$K = \tfrac{1}{3} n \bar{c} \lambda (c_e), \tag{9.56}$$

whilst the electrical conductivity, from equation (9.48), may be written

$$\kappa = \frac{\lambda e^2 n}{2m\bar{c}}. \tag{9.48}$$

The ratio is

$$\frac{K}{\kappa} = \frac{2}{3} \frac{m\bar{c}^2}{e^2} (c_e). \tag{9.57}$$

(At this point if we assume the electrons behave like a perfect gas $m\bar{c}^2$ is $3kT$ and $c_e = \tfrac{3}{2} k$; we then obtain precisely the classical Drude result given in equation (9.52).)

The quantity $\bar{c}$, the velocity of those electrons in the Fermi distribution which take part in the thermal transport process, is of order

$$\bar{c}^2 \approx \frac{E_F}{m}, \tag{9.58}$$

so that equation (9.57) becomes, after substituting for $c_e$ from equation (9.55), and for $\bar{c}^2$ from equation (9.58),

$$\frac{K}{\kappa} \approx \frac{E_F}{e^2} (c_e) \approx \frac{k^2 T}{e^2}. \tag{9.59}$$

A more rigorous treatment gives a constant $\tfrac{1}{3} \pi^2$ in front of this expression so that it is very nearly the same as equation (9.52) obtained from the simple Drude model.

The conclusions one derives from this, however, expose another issue of very great importance. The value of $E_F$ of about 2 eV implies that the electrons have velocities (see equation (9.58)) about 10 times greater than the free electron-gas velocity. To explain the observed conductivities one must assume a mean free path $\lambda$ of the order of 50–100 nm. On the Drude model $\lambda$ has a value comparable to the distance between the positive ions in the lattice; this seems reasonable. A value of 100 atomic spacings or so seems strange. The explanation lies in the fact that electrons travelling through a perfect lattice experience no resistance whatsoever. If we treat the electrons as waves it is possible to show that electrons scattered by the lattice recombine to give a wave of undiminished amplitude; in other words there is no attenuation of the wave so that the resistance is zero. At normal temperatures the lattice vibrations render the lattice irregular so that the electron waves are scattered in an irregular way. It is this which accounts for the observed finite resistance. Any factor which increases the irregularity of the lattice increases the resistance. This is why the resistance increases with temperature and with impurity content.

# 10

## *The liquid state*

The main characteristics of the gaseous state are random, translational motion of the molecules and relatively little molecular interaction. In the solid state the main characteristics are fixed sites for the individual atoms or molecules and very little translational movement. Both these states are well understood. By contrast the liquid state, in some ways, has 'no right to exist'. The molecular interaction must be quite strong since a given quantity of liquid, like a solid (and unlike a gas) occupies a definite volume. On the other hand the molecules have so much freedom that a liquid, unlike a solid, flows very readily and easily takes up the shape of the vessel it occupies.

The gaseous state has been thoroughly investigated and most of its problems have been satisfactorily resolved. The solid state is still being studied in hundreds of physics laboratories all over the world. By contrast the liquid state is a neglected step-child of the physical scientists. This is partly because the liquid state raises a number of very difficult theoretical problems. It is also partly a matter of fashion dictated by the economic value and engineering uses of solids as structural materials, electrical devices, etc. But for chemists, physical chemists and especially biologists, the liquid state is possibly of greater importance than the solid state. We may well expect, we should certainly hope for, marked advances in our understanding of the liquid state in the next two or three decades.

There are basically three approaches to the liquid state. The first is to treat it as a modified gas; the second is to treat it as a modified solid; the third is to treat it *sui generis* – this is the most difficult of all.

### 10.1 The liquid as a modified gas

A liquid in static equilibrium cannot withstand a shear stress. If we compress a column of liquid with a piston the situation is not like that of a solid where a uniaxial pressure or tension produces a shear stress in the medium. Unlike a solid the liquid must be held in a container; the walls exert pressures on the liquid in such a way that no resultant shear stress remains. This is equivalent to saying that the only stress the liquid can experience is a hydrostatic pressure or tension. It is for this reason that, at any point in a stationary liquid, the pressure is the same in all directions. In this sense a liquid closely resembles a gas and indeed both states are sometimes designated by the single generic term 'fluid'. In terms of resistance to pressure it therefore seems reasonable to consider how far the gas laws can be applied to explain the behaviour of liquids. Naturally we would not expect the ideal gas laws to be applicable but we might expect the imperfect gas laws to have some relevance.

We start by considering a liquid as a limiting case of a van der Waals gas. We ask what meaning may be attached to the van der Waals constants in the equation

$$\left(P + \frac{a}{V^2}\right)(V - b) = RT.$$

(a) The quantity b. In a gas $b = 4v_m$ and at the critical volume $V_c = 3b$ so that the quantity $(V_c - b)$ is still positive and meaningful. At temperatures well below the critical temperature the density of the liquid increases, the volume occupied by the liquid is only perhaps $(1.5$ to $2)v_m$ so that $(V - b)$ is negative unless a different correction value is given to $b$.

(b) The quantity $a/V^2$. So long as the molar volume is large $a/V^2$ is small and is meaningful as a correction term. However, when $V$ is comparable with or less than $b$ the term $a/V^2$ has a value of the order of $1000$–$10\,000$ atmospheres. This internal pressure is so large compared with the atmospheric pressure that it cannot be treated as a correction factor. This is, of course, consistent with the fact that the external pressure $P$ is not an important factor in determining the volume occupied by a liquid.

(c) The tensile strength of a liquid. In what follows we shall show that for low-temperature isotherms a liquid, according to van der Waals' equation, can actually withstand negative pressures. If we apply an increasing negative pressure (hydrostatic tension) the liquid will expand until at some critical value vapour will begin to form within the liquid, that is to say the liquid begins to cavitate or pull apart. The pressure at which this occurs provides, therefore, a measure of the tensile strength of the liquid.

Following Temperley (1947)† we use the reduced equation of state and write

$$\left(\pi + \frac{3}{\phi^2}\right)(3\phi - 1) = 8\theta, \tag{10.1}$$

where $\pi$, $\phi$ and $\theta$ are the reduced pressure, volume and temperature respectively. We may expand the liquid from D to B (see figure 77) by applying a hydrostatic tension. At B the liquid would become a vapour, i.e. it would pull apart. The pressure at which this occurs may be obtained by writing equation (10.1) as

$$\pi = \frac{8\theta}{(3\phi - 1)} - \frac{3}{\phi^2}, \tag{10.2}$$

and determining the minimum at B by putting $\partial \pi / \partial \phi = 0$. This gives

$$\frac{-3 \times 8\theta}{(3\phi - 1)^2} + \left(\frac{6}{\phi^3}\right) = 0.$$

This defines the value of $\phi$ at B; we call this $\phi_m$. Then

$$4\phi_m^3 \theta = (3\phi_m - 1)^2. \tag{10.3}$$

Figure 77. The tensile strength of a liquid treated as an extreme case of a van der Waals fluid. Using the reduced equation of state it is shown that the liquid fails when a negative pressure $\pi_m$ is applied.

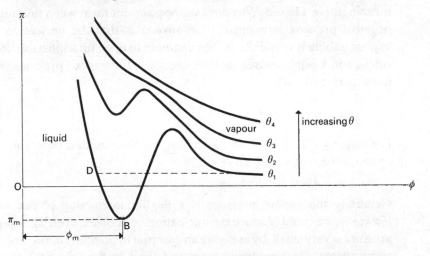

† H. N. V. Temperley, *Proceedings of the Physical Society*, vol. 59 (1947) p. 199. See also A. H. Cottrell, *The Mechanical Properties of Matter* (Wiley, 1964).

If we insert this in equation (10.2) we obtain the corresponding value of $\pi$ which we call $\pi_m$.

$$\pi_m = \frac{3\phi_m - 2}{\phi_m^3}. \tag{10.4}$$

We see from this equation that $\pi_m = 0$ when $\phi_m = \frac{2}{3}$. This is the 'last' isotherm for which the liquid has no tensile strength; it occurs when $\phi = \frac{2}{3}$ or $V = \frac{2}{3}V_c$. If the liquid is less condensed than this it can exist only if a positive pressure is applied.

At lower temperatures $-\pi_m$ becomes larger. It reaches its maximum (negative) value, as equation (10.3) shows, when $\theta = 0$, i.e. when $3\phi = 1$. Putting $\phi = \frac{1}{3}$ in equation (10.4) we obtain as the maximum value

$$\pi_m = \frac{1-2}{(\frac{1}{3})^3} = -27. \tag{10.5}$$

The limiting hydrostatic tension is then

$$P = -27P_c.$$

For water $P_c = 218$ atmospheres, so that

$$P = -5900 \text{ atmospheres.} \tag{10.6}$$

(This is comparable with, though appreciably larger than the internal pressure quoted on p. 126). Before we leave this calculation we note that $\theta = 0$ corresponds to absolute zero temperature and $\phi = \frac{1}{3}$ corresponds to $V = \frac{1}{3}V_c = b$, i.e. the molecules are packed as close as the van der Waals model will allow. We may ask how such high negative pressures can be meaningful in a liquid. Why does the vapour not form when the smallest negative pressure is applied? The answer is that the nucleation of a vapour bubble is very difficult. For example to open up a spherical hole of radius $r$ in a liquid of free surface energy $\gamma$, the vapour pressure would have to exceed

$$P = \frac{2\gamma}{r}. \tag{10.7}$$

For water $\gamma = 72$ mJ m$^{-2}$ and assuming $r$ is the size of a water molecule (=about 0.3 nm) we obtain

$$P = 4800 \text{ atmospheres.} \tag{10.8}$$

Generally the vapour pressure is a negligible fraction of this value. However, we could produce the nucleation of bubbles, even if the vapour pressure is very small, by applying an external hydrostatic tension of 4800 atmospheres. Cavities would form and the liquid would then rupture. This tension is very close to that calculated in equation (10.6).

Readers may note that we could, in fact, estimate the breaking strength of the liquid simply in terms of the surface energy. Consider a liquid column of unit cross-section. If we pull it apart we create two unit areas of surface, so that the work done is $2\gamma$. The intermolecular forces are short range and if they become negligible for a distance greater than say $x$, the force $F$ to pull the column apart will be of order $F \times x = 2\gamma$ or $F = 2\gamma/x$. If $x$ is about 0.5 nm we again obtain as the breaking stress (force per unit area) a value close to that given in equations (10.6) and (10.8). Exact agreement is not to be expected; the main point we wish to make is that because of intermolecular forces liquids can, in theory, withstand large negative pressures. In practice, of course, liquids are generally much weaker because of dissolved gases which readily nucleate cavities as the negative tension is applied. This is particularly marked at the walls of the vessel where small amounts of gas may be trapped in the surface irregularities. Carefully outgassed liquids in smooth-walled containers can, however, possess tensile strengths up to one-tenth of the values calculated above.

### 10.2 **The structure of liquids: the 'radial distribution function'**

Before turning to a consideration of a liquid as a modified solid we mention one characteristic feature of a liquid which distinguishes it in a very basic way from a gas. A gas has no structure. If we study the diffraction of neutrons, electrons or X-rays by a monatomic gas we find that there is no coherent diffraction; the gas behaves like a diffraction grating in which the spacings between the lines are random. (For diatomic molecules one obtains diffraction maxima corresponding to the mean distance between the atoms in the molecule.) If, however, a monatomic gas is liquefied it shows marked coherent diffraction; the maxima are not sharp as with a solid crystal, but are diffuse.

The information obtained by such diffraction experiments is presented most graphically by means of the 'radial distribution function' which may be derived in the following way. We choose a molecule at random within the liquid and draw a series of concentric spheres around it such that the volume interval between two neighbouring spheres is a constant. The distribution function is the average number of molecules contained between each spherical shell. It is evident that the radial distribution function is a measure of the average density as a function of distance from some arbitrary origin. A typical result from X-ray diffraction is given in figure 78 and similar results are obtained from the diffraction of neutrons. As a matter of interest the radial distribution function for the crystalline solid is also shown; this brings out the marked difference between the liquid and the solid state. The solid behaves like a diffraction grating containing a very large number of regularly spaced lines: the liquid resembles the diffraction of light by a grating containing only a few lines. This suggests that there is short-range order in a liquid; any small group of atoms or molecules is in fairly regular array and resembles a crystal lattice but it is just a little out of register in a random way so that, as soon as one moves a short distance from the group, the neighbouring atoms (or molecules) no longer possess the long-range order characteristic of the solid crystal.

The contrast between the liquid and solid states may be expressed in another way.† Neutron diffraction experiments suggest that a molecule in

Figure 78. Radial distribution function plotted against radius measured from an arbitrary atom chosen as origin. (*a*) Liquid mercury (*b*) solid mercury (from results given by A. Guinier, *X-ray Diffraction in Crystals, Imperfect Crystals and Amorphous Bodies* (Freeman, 1963) p. 70, figure 3.9). For the liquid, the vertical ordinate can be normalized such that for large distances the radial distribution function tends to unity (uniform density). For the solid this is not possible since in principle each peak is infinitely narrow and infinitely high. Instead the vertical ordinate has been drawn proportional to the number of atoms at the specified radial distance. It is seen that with the liquid the structure is lost beyond distances of 1 nm; with the solid it persists indefinitely.

liquid Hg (normalized height units)
solid Hg (arbitrary height units)

† See J. D. Bernal, *Scientific American*, no. 203 (1960), pp. 124–34; and *Proceedings of the Royal Society*, A280 (1962), pp. 299–322.

a liquid has time to vibrate 10 to 100 times before the local structure changes. During that time the structure of a liquid is physically, though not geometrically, similar to that of a crystal. The difference, however, is crucial. For example the solid crystal, because of its high degree of order, admits only a limited degree of variation of composition. Crystallization as a means of purifying substances is evidence of the particular architecture that marks each crystalline phase. By contrast, liquids mix much more readily either with other liquids or with solids or with gases (though of course they do not mix as readily as gases which are always perfectly miscible with one another). We may describe this situation by saying that 'atoms in a liquid do not occupy such closely specified positions as they do in solids. They have more elbow room, and they are not so particular about their partners.'

Nevertheless the structural resemblance to the solid over limited regions appears to be indisputable. This conclusion is supported from a completely different type of measurement. It is found that when metals melt their electrical resistance increases by a factor of only about two. As mentioned in a previous chapter the resistance is primarily due to non-coherent scattering of the electrons by the metal ions. This implies that, to the conductance electrons, the metal ions, in the liquid, appear to be oscillating about regular sites, very much as in the solid state.

There are two main differences between the solid and liquid state. First, as discussed above, there is the absence of long-range order in liquids: in this sense they closely resemble amorphous solids, glasses and polymers. It is significant that these materials can flow viscously like liquids although, as pointed out in Chapter 6, their viscosity is generally enormous compared with that of true liquids. On the other hand, with crystalline solids flow or 'creep' can occur, but it takes place extremely slowly and through an entirely different mechanism – through the movement of dislocations, the behaviour of which is determined by the highly ordered structure of the material. This brings us to the second major difference: compared with the molecules in the solid state (both crystalline and amorphous), the molecules in liquids have very much greater mobility. They can swap places or shuffle around to other positions with relative ease.

### 10.3 The liquid as a modified solid

When solids melt they expand generally by 5–15 per cent in volume. One way of describing this is to say that there is an increase in free volume between the constituent molecules or atoms. Another is as follows (cf. E. A. Moelwyn-Hughes, private communication).† If one thinks of a typical crystalline solid for which each molecule, or atom, has say 8 or 12 nearest neighbours, we may regard the liquid state as a solid from which one nearest neighbour has been subtracted. This leaves the material with the right amount of volume expansion and a reasonable degree of short-range order. What is the nature of this disorder? The X-ray diffraction evidence suggests that the mean intermolecular distance does not change as much as would be expected from the bulk volume expansion. For example when solid argon melts the mean argon–argon spacing increases by only about 1 per cent, whereas the volume change would lead us to expect a mean increase in separation of about 5 per cent. Such evidence may be used to justify the view that the melting process consists mainly in producing 'holes' in a structure which otherwise closely resembles the solid structure. This idea was first developed by J. G. Kirkwood and constitutes the simplest form of the 'hole-theory' of liquids. As we shall see later the modified 'hole-theory' introduced by H. Eyring provides a very convenient model for the study of viscous flow in liquids.

Attempts have been made to treat the disordered structure and the increase in free volume in the liquid state in terms familiar to the solid-state physicist. One approach is to concentrate on the role of vacancies or holes (A.L.G. Rees, 1966, private communication). As the temperature rises the concentration of vacancies increases. So long as vacancies are individual holes the behaviour is still that of a solid. As soon, however, as two or three vacancies coalesce a whole aggregate of molecules surrounding the hole becomes relatively mobile. The material is now very much more like a liquid. One advantage of this approach is that it stresses the cooperative nature of the melting process.

### 10.4 **The liquid state *sui generis***

The most difficult approach is to tackle the liquid state without directly referring to the gaseous or solid state. The simplest method is that due to J. D. Bernal and his colleagues in England and to G. D. Scott in Canada. It may be referred to as a random-packing model. Recognizing the disordered but relatively close-packed nature of the liquid structure Bernal took a large number of plasticine spheres, chalked them to stop them sticking, placed them in a football bladder, removed the air to avoid bubbles and them squeezed them together until they filled the whole of

† See also E. A. Moelwyn-Hughes, *Physical Chemistry* (Pergamon Press, 1960).

the space. On examining the aggregate it was found that the spheres had become polyhedra of various irregular shapes. The most common number of faces was thirteen (this was not the true coordination number since the centres of some of the neighbours were more than the average distance apart). More striking was the observation that the most common number of sides to a face was five. This is of great significance for the following reason. In crystallography molecules can be arranged with two-fold, three-fold, four-fold, or six-fold symmetry, never with five-fold symmetry. This is because 'one cannot form regular patterns with five-fold symmetry that will fill space solidly and extend indefinitely in three dimensions. It is like trying to pave a floor with five-sided tiles'. This is illustrated in figure 79 and shows how this type of arrangement would explain the presence of short-range order and the absence of long-range order.

In a later study Bernal emphasized the point that the attractive forces between molecules play little part in influencing the packing which occurs in the condensed state: the packing is almost entirely determined by the repulsive forces, i.e. the molecules behave very much like hard spheres. As Bernal remarks, the key word (originally used in a different sense) is 'the one used by Humpty Dumpty in *Alice Through the Looking Glass*, impenetrability'. We are familiar with this in the solid state where, for practically all crystalline structures which are not dominated by covalent or other directed bondings, the arrangement is rather like that arising from packing the molecules as close as their repulsive forces will allow. For simple atoms, indeed, the crystal structure is like that of close-packed spheres. Bernal therefore considered the random packing, not of deformable plasticine spheres as in the earlier work, but the packing of a

Figure 79. The difference between five-fold symmetry and six-fold symmetry, after Bernal.

5-fold                                    6-fold

large number of hard steel balls. After introducing black paint and letting it harden the aggregate could be broken open and studied. In this way the number of contacts and near contacts could be determined. The model gives a coordination number varying between four and eleven, and Bernal considers this variation in contact number from molecule to molecule to be the most significant feature of the irregular liquid structure. The packing very adequately reproduces the radial distribution function for liquid argon derived from neutron diffraction experiments (see figure 80), though this is probably not a very critical test. Further, the model does not allow vacancies between neighbours to be large enough to admit a third molecule.

The Bernal model aims to provide essentially an instantaneous snapshot (in say $10^{-15}$ s) of the structure of a liquid. If we now add to this the reality of thermal motion we see that the molecules are continuously changing the number of nearest neighbours, continuously shifting and shuffling around, and occasionally are able to allow one molecule to squeeze its way between neighbours without leaving a vacant hole. This aspect of the model may be open to criticism. For example with a solid, because it is rigid, vacant sites or holes can exist in the lattice. The concentration of holes increases as the temperature is raised and by the time the solid reaches its melting point (but is not yet melted) about one site in $10^4$ or $10^5$ is vacant. Once melting occurs there is a further increase in volume, but according to Bernal the free space is more uniformly spread out. The X-ray evidence quoted above suggests that this is not

Figure 80. Radial distribution function for liquid argon plotted against distance from centre in units of sphere diameter. — results from neutron diffraction; ●▽ calculated from random packing model (● Bernal, ▽ Scott).

possible, even at temperatures just above the melting point; in any case the thermal expansion of liquids is rather large so that at somewhat higher temperatures there must be vacancies or holes. Indeed, as Bernal himself points out, it is this increase in holes and the consequent drop in average coordination number that marks the transition from the coherent liquid phase to the incoherent gas phase (see figure 81). Bernal suggests that this probably occurs when the coordination number has fallen to an average value of three or four. Since the coordination number in the solid state (or in the liquid state near the freezing point) is about twelve, this means that the density of a gas at its critical point is between one-quarter and one-third of the density of the most highly condensed state. This is in fairly good agreement with observation for most *normal* liquids.

The Bernal approach is essentially that of a crystallographer; his model is a static one and concentrates on structural features. A far more analytical and difficult approach is that due to M. Born and H. S. Green. They carried out an extensive study of the mutual potential between molecules in the liquid state. They treat the molecules first in pairs, then in triplets and so on. The treatment is extremely complicated and the results are not easy to apply. In a sense their approach approximates to the theory of dense gases. A calculation is made of sets of coordinates of random collection of hard spheres. As the density of spheres is increased the behaviour passes from that of a gas to that of a dense gas and then to some other state. In this condition, by an iterative process, the coordinate positions can be repeatedly perturbed until a state of minimum energy is obtained. This corresponds over a critical density range to the liquid state. The computation here is formidable even for only 100 spheres.

Figure 81. The transitions from the close-packed coherent liquid phase to the expanded, associated and finally free (incoherent gas) phase, as proposed by Bernal.

Yet another approach is that associated with the name of B. J. Alder; it is sometimes referred to as the method of molecular dynamics. It involves elaborate computations but the basic concept is simple. The molecules are treated as hard smooth perfectly elastic spheres, although as a later refinement an attractive square-well potential with a rigid core has been used. The molecules are placed in a box and started off with equal speeds in random directions and their subsequent motion determined by applying Newton's laws of motion. The boundary conditions chosen are such that a particle which passes out of one side of the box re-enters with the same velocity through the opposite side. In this way the number of molecules in the box and the total energy remain constant. The collisions within the system are analysed by a computer. It is found that after relatively few collisions the velocity distribution becomes very nearly Maxwellian. The total energy is determined by the initial velocities and of course remains constant; it determines the effective temperature of the system.

The behaviour depends on the velocity and on the packing of the particles. At one extreme the particles oscillate about an equilibrium position; this corresponds to the solid state. At the other extreme the particles move around throughout the box with relative freedom; this corresponds to the vapour regime. At some intermediate stage the

Figure 82. End face of box showing calculated particle trajectories for 32 hard elastic spheres (zero attractive forces) for a packing $V/V_0 = 1.525$. (*a*) Behaviour resembling a solid. (*b*) Behaviour resembling a liquid. The particles spend most of their time oscillating within a *cage* of other particles, but are also able to swap places. See particles A and B; C, D and E.

(a)  (b)

particles vibrate about equilibrium positions but are also able to change places and slowly diffuse from one region to another. This corresponds to the liquid state.

We quote some typical results obtained by Alder and Wainwright for 32 hard smooth particles in a box.† They were able to map out a view of the calculated particle trajectories as seen from one face of the box. If $V$ is the volume occupied by the particles and $V_0$ the volume they would occupy in a close-packed situation, they observed a very interesting instability for a ratio $V/V_0 = 1.525$. For one group of collisions the particle trajectories appear as shown in figure 82($a$) and calculation shows that the pressure on the walls is small; clearly this represents the solid state. In other groups of collisions the particle trajectories appear as in figure 82($b$) and calculations show that the pressure on the walls is large. This is interpreted as the liquid phase. It is interesting to note that in the 'liquid' state the particles oscillate about a mean position but occasionally swop places (see figure 82($b$), particles A and B; particles C, D and E). The situation shown in these figures refers to only 32 particles in the box. Apparently the system is not sufficiently large for the solid and liquid phases to exist in equilibrium: the system is in just that density range where part of the time it is in the liquid phase and part of the time in the solid phase.

Although the results reproduced in figure 82 are very striking, some reserve is necessary since an assembly of particles which show no attraction for one another can hardly be considered as a very realistic model of a liquid or a solid. However, from the point of view of packing, the structure scarcely depends on intermolecular forces (see Bernal above).

If the molecules are less closely packed the results obtained are as shown in figure 83. Figure 83($a$) shows the potential for particles with a square-well attractive field whilst figure 83($b$) shows the particle trajectories for 108 such particles in a box; the behaviour evidently represents the liquid–vapour region. The Alder model brings out very clearly the dynamic features of the condensed state and the similarities and differences between the solid and liquid phases.

### 10.5 Ice and water

We introduce here a short digression on the relation between ice and water. Water, which is the basis of all our ideas of liquidity, is in a sense an atypical liquid. The forces between the molecules are very

† B. J. Alder & T. E. Wainwright, *Nuova Cimento*, vol. 9 (1958), sup. 1, p. 16, and *Journal of Chemical Physics*, vol. 31 (1959), p. 459.

strongly directional; they arise from the interaction between electron-deficient hydrogen atoms in one molecule and electron-rich oxygen atoms in another. In the solid state this hydrogen bonding leads to a very open structure in which every water molecule is surrounded symmetrically (in the form of a regular tetrahedron) by four other water molecules. Consequently in ice the coordination number is four. As the temperature is raised the coordination number remains constant but the mean distance between the molecules increases. There is thus a small decrease in density with increasing temperature as with all 'normal' solids. At the melting point, however, a radical change occurs. The thermal energy is sufficient to destroy the regular bonding and the coordination number actually *increases*. At 0 °C the average coordination number appears to be between 4.5 and 5; with rising temperature this probably increases but soon reaches a constant value. This increase on its own must lead to a corresponding *rise* in density. However, an increase in temperature also leads to an overall increase in the mean distance between the molecules (thermal expansion). This expansion, on its own, would produce a *decrease* in density with increasing temperature. When combined with the change in coordination number it results in a 10 per cent jump in density on melting, a maximum density at 4 °C followed by a steady decrease in density at temperatures above about 10 °C. The behaviour is

Figure 83. (*a*) Potential energy diagram of spheres possessing a square-well attractive field and a hard elastic core. The repulsive core has an effective diameter $\sigma_1$. The attractive forces operate at an effective separation of $\sigma_2$. (*b*) End face of box showing calculated trajectories of 108 such spheres. The behaviour of the particles resembles the vapour phase.

(a)

(b)

shown schematically in figure 84. The details are by no means definite but the general pattern is reasonably valid. Another way of looking at this is in terms of the 'mixture-models' of water (e.g. H. S. Frank models). These stress the idea that water may not simply be a random or an ordered arrangement of individual molecules with uniform characteristics and bonding. On the contrary water may consist of two or more different molecular species, each species being structurally organized in its own distinctive way, with its own degree of coordination. The effect of temperature on this model is to increase the proportion of one species at the expense of the other. Whichever approach is adopted, the main point brought out by this discussion is the immense importance of the coordination number in determining the detailed properties of liquids.

## 10.6 General approach

In what follows we shall treat the liquid as a modified solid, particularly when we are concerned with energies. When we consider

Figure 84. The melting of ice. Schematic diagram showing (*a*) the change in coordination number, (*b*) the change in molecular separation, (*c*) the corresponding change in density as ice is heated from $-5\,°C$ to $+15\,°C$.

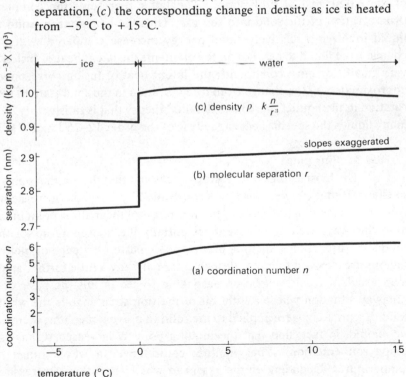

temperature (°C)

flow we shall treat the liquid essentially as a solid, the molecules of which have a fair degree of mobility because of the presence of holes or free volume. This is the Eyring theory of viscous flow in a liquid; it differs from the Bernal theory in concentrating on thermal fluctuations rather than on the structural nature of the liquid state. It is much closer to the Alder model of the liquid state.

Before leaving this discussion we may make the following point. In the gaseous state the molecules are, more or less, completely independent. In the solid state they interact but the interaction, at least as far as crystalline solids are concerned, is capable of analysis precisely because of the long-range order which obtains. Solids are indeed an example of highly ordered cooperative interaction. By contrast the behaviour of a liquid is the result of cooperative interaction within a system of random order, or, at best, low-range order. It is this feature which makes the theory of liquids so difficult.

### 10.7 Latent heat of fusion

The latent heat of fusion is between one-tenth and one-thirtieth of the latent heat of sublimation. The latter represents the complete disintegration of the solid into the gas. Thus the change from solid to liquid involves a relatively small energy increase compared with the change from liquid to gas. For all the elements the latent heat of fusion is very small; for most compounds the latent heat of fusion corresponds approximately to the change from free vibration in the solid state to free rotation in the liquid. There is no detailed theory that is satisfactory. For many liquids the specific heat capacity is of the order $12–25$ J mol$^{-1}$ K$^{-1}$.

### 10.8 Melting point

The most astonishing feature of fusion is that the melting point is so sharp. If melting were always accompanied by a volume increase we could understand (if not explain) the sharpness of the melting point in the following way. Melting represents essentially the 'loosening-up' of the solid structure. If local melting occurs molecules on the edge of melted regions are exposed to molecules which are already a little further apart than usual. It would not be unreasonable to see in this the basis of a runaway situation where a little bit of melting soon engulfs the whole solid. If a pressure were applied to the solid to prevent a volume increase we should, in fact, find that the melting point is increased (Clausius–Clapeyron equation). Thus melting could cover a whole range of temperatures depending on the extent to which the volume increase is inhibited; only if the solid were completely free to expand would melting occur at a single sharply defined temperature. This 'explanation' would, of course, be a little embarrassing if we attempted to apply it to the melting of ice since here melting is accompanied by a decrease in volume.

As we saw earlier it is not possible to explain melting in terms of our potential energy curve, since this can only describe basically the solid state or the gaseous state. However, Lindemann (1910) suggested that one could obtain a good idea of the melting point from the potential energy curve by assuming that melting occurs when the amplitude of vibration exceeds a critical fraction of the atomic spacing. At this stage numerous collisions occur between neighbours and the crystal structure is destroyed.

Let $x_0$ be the vibration-amplitude when melting occurs. If $f$ is the force constant between one atom (or molecule) and its neighbour and the motion is simple harmonic, the mean total energy during vibration will be

$$u_v = \tfrac{1}{2}fx_0^2. \tag{10.9}$$

Since melting points are generally high we may assume the classical value for the internal energy, namely $kT$ per atom or molecule for each principal direction of vibration. Thus at the melting point we write

$$u_v = kT_m = \tfrac{1}{2}fx_0^2. \tag{10.10}$$

But as we saw in an earlier chapter, for a simple cubic structure, $f = Ea$, where $a$ is the atomic (or molecular) spacing and $E$ is Young's modulus of the solid.

Hence $kT_m = \tfrac{1}{2}Eax_0^2.$

We assume that melting occurs when $x_0$ is some fraction of the atomic spacing $a$. Let $x_0$ be $\beta a$ where $0 < \beta < 1$.

$$T_m = \frac{Ea^3\beta^2}{2k} = \frac{Ea^3N\beta^2}{2R} = \frac{EV\beta^2}{2R},$$

where $V =$ molecular volume, assuming a cubic structure.

$$T_m = \frac{E}{2\rho}\frac{M}{R}\beta^2, \tag{10.11}$$

where $\rho =$ density of the solid in $kg\,m^{-3}$, $M$ is the molar mass in kg, and $R = 8.3\,J\,mol^{-1}\,K^{-1}$.

Some typical results are given in Table 28 where it is assumed that $\beta^2$ has the value of $\frac{1}{50}$, i.e. $x_0 \approx \frac{1}{7}a$. Note that this model has assumed simple cubic packing (so that $Na^3 = V$) and $\beta$ a constant fractional amplitude of vibration at melting. Considering the crudity of the model and the range of materials to which it has been applied the agreement is surprisingly good.

## 10.9 Vapour pressure

### 10.9.1 Molecular interpretation of vapour pressure

Suppose we place a quantity of liquid in a sealed container and evacuate the air above it. Some of the liquid will evaporate since a certain fraction of the molecules will have enough excess thermal energy to escape from the attraction of neighbours. Some of these will return and condense. Equilibrium will be reached when the rate of evaporation equals the rate of condensation. The pressure exerted by the vapour under these conditions is known as the saturation vapour pressure or, more usually, the vapour pressure. It increases as the temperature is raised. The equilibrium vapour pressure will be almost identical in the presence of air or another gas (see, however, the section below on the effect of external pressure), since the vapour and the alien gas exert their partial pressures independently.

If the pressure above the liquid is set at some specified level, for example 1 atmosphere, the vapour pressure may be increased by raising the temperature of the liquid, until it is equal to the external pressure. At this point the liquid can evaporate freely: bubbles of vapour form copiously within the bulk of the liquid and the liquid is said to be boiling. The boiling point is thus the temperature at which the vapour pressure equals the external pressure – usually one atmosphere. If the external pressure is reduced the boiling point is lowered. It is a matter of experience that the boiling point on the absolute scale, at atmospheric pressure, is approximately two-thirds of the critical temperature, but there is no simple explanation for this.

Table 28. Melting point $T_m(K)$ of some solids, assuming $\beta^2 = 1/50$

| Solid | $E$ (N m$^{-2}$) | $\rho$ (kg m$^{-3}$) | $M$ (kg) | $T_m$ (K) Calculated | Observed |
|---|---|---|---|---|---|
| Lead | $1.6 \times 10^{10}$ | $11.3 \times 10^3$ | 0.207 | 400 | 600 |
| Silver | $8.3 \times 10^{10}$ | $10.5 \times 10^3$ | 0.108 | 1100 | 1270 |
| Iron | $21.2 \times 10^{10}$ | $7.9 \times 10^3$ | 0.056 | 1800 | 1800 |
| Tungsten | $36.0 \times 10^{10}$ | $19.3 \times 10^3$ | 0.184 | 4200 | 3650 |
| Sodium chloride | $4.0 \times 10^{10}$ | $2.16 \times 10^3$ | 0.057 | 1200 | 1070 |
| Quartz | $7.0 \times 10^{10}$ | $2.6 \times 10^3$ | 0.060 | 1900 | 2000 |

The molecular interpretation of vapour pressure is simple. The molecules in the liquid have a wide range of thermal energies. The only molecules which can escape are those whose thermal energy is sufficient to overcome the attraction of neighbouring molecules. If the latent heat of vaporization per mole is $L$, the energy for a molecule to escape is

$$\varepsilon = \frac{L}{N_A}.\qquad(10.12)$$

It follows that the average potential energy of the molecules in the vapour phase is greater than the mean value in the liquid phase by an amount $\varepsilon$. Consequently if $n_V$ and $n_L$ represent the number of molecules per unit volume in the vapour and liquid phases respectively and if the Boltzmann distribution holds for the liquid as well as for the vapour, we have

$$\frac{n_V}{n_L} = \exp\left(-\frac{\varepsilon}{kT}\right) = \exp\left(-\frac{L}{N_A kT}\right) = \exp\left(-\frac{L}{RT}\right).\qquad(10.13)$$

As the temperature is raised the ratio $n_V/n_L$ increases but $n_L$ scarcely changes with temperature. Since the vapour pressure $p$ is directly proportional to the number of molecules per unit volume $n_V$ in the vapour phase and to the temperature $T$ we may re-write equation (10.13) as

$$p = BT \exp\left(-\frac{L}{RT}\right).\qquad(10.14)$$

We shall find in many expressions of this type that the variation of the exponential term with temperature is far greater than the variation of the pre-exponential term. For example if $T$ is increased from 0 °C to 30 °C the pre-exponential term increases by a factor of 1.1, whereas, for say water, the exponential term increases by a *factor* of 8. Since equation (10.14) is, in any case, based on a very simple model we may simplify it further by ignoring the pre-exponential term in $T$. We may then write

$$p = A \exp\left(-\frac{L}{RT}\right),\qquad(10.15)$$

where $A$ is a suitable constant. This relation is almost exactly correct.

We may approach the vapour pressure relation in another way. Suppose we consider the liquid to resemble a very dense gas containing $n_L$ molecules per m³. The number per m³ with velocities, normal to the surface, between $u$ and $u + du$ is given by equation (4.41) in Chapter 4. The number striking each m² of surface per second is then

$$u \, dn = n_L \left(\frac{m}{2\pi kT}\right)^{\frac{1}{2}} \int_0^\infty \exp\left(-\frac{mu^2}{2kT}\right) u \, du.$$

If we now specify that of these molecules only those with energy greater than $\varepsilon$ (where $\varepsilon = L/N_A$) can escape, the number per second crossing each $m^2$ of surface into the vapour phase is the same integral as above but the limits are from $mu^2/2 = \varepsilon$ to $\infty$. This gives us the number escaping per $m^2$ per second,

$$n_L\left(\frac{kT}{2\pi m}\right)^{\frac{1}{2}} \exp\left(-\frac{\varepsilon}{kT}\right).$$

But if $n_V$ represents the number of molecules per $m^3$ in the vapour phase, the number striking each $m^2$ of surface per second will be (see Chapter 4, equation (4.41))

$$n_V\left(\frac{kT}{2\pi m}\right)^{\frac{1}{2}}.$$

If all those molecules striking the liquid surface are adsorbed (this is the most dubious part of the treatment) then equilibrium occurs when the number escaping from the liquid exactly equals the number arriving from the vapour phase, i.e.

$$n_L \exp\left(-\frac{\varepsilon}{kT}\right) = n_V.$$

This is identical with equation (10.13).

### 10.9.2 *Effect of temperature on vapour pressure*
We use equation (10.13) to calculate the effect of temperature on vapour pressure. Since

$$p = A \exp\left(-\frac{L}{RT}\right),$$

we have by differentiating

$$\frac{dp}{dT} = A\frac{L}{RT^2}\exp\left(-\frac{L}{RT}\right) = \frac{L}{RT^2}\times p = \frac{L}{RT}\times\frac{p}{T}. \tag{10.16}$$

If the vapour behaves like a perfect gas and $V$ is the volume occupied by one mole of vapour at $T$ we have $pV = RT$, so that

$$p = \frac{RT}{V}.$$

Equation (10.16) becomes

$$\frac{dp}{dT} = \frac{L}{T}\times\frac{1}{V}. \tag{10.17}$$

The correct relation, according to thermodynamics, is

$$\frac{dp}{dT} = \frac{L}{T} \frac{1}{(V - V_L)},$$ (10.18)

where $V_L$ is the volume occupied by one mole of the liquid. This is known as the Clausius–Clapeyron equation. We may note that a plot of $\ln p$ against $T^{-1}$ should give a straight line of slope equal to $L/R$. This is indeed found to be so.

### 10.9.3 *Effect of external pressure on vapour pressure*

If we apply an external pressure, for example by compressing with an inert gas, how is the vapour pressure $p$ of the liquid affected? The answer is surprising; $p$ is increased. This may be explained in terms of figure 85 which represents an idealized experiment. The top horizontal limb of the tube abcd contains a membrane permeable only to the vapour above the liquid AB. When the space above the liquid contains only vapour, A is level with B. If now an inert gas is pumped into the right-hand limb at a pressure $\pi$, since it cannot pass through the membrane, the level A rises to A' and B falls to B'. If $h$ is the difference in height, clearly

$$\pi = h\rho_L g.$$ (10.19)

Since the membrane is permeable to the vapour, the vapour over A' must be in equilibrium with the vapour over B'. This implies that the vapour pressure at B' is greater than over A' by $h\rho_V g$. The increase in vapour

Figure 85. Effect of external pressure on the vapour pressure of a liquid. (*a*) Initial condition – the membrane is permeable only to the vapour. (*b*) Condition when external pressure is applied. (*c*) Alternative model.

(a)    (b)    (c)

pressure is thus

$$\Delta p = h\rho_V g = \pi \frac{\rho_V}{\rho_L}. \tag{10.20}$$

Another way of understanding this is to consider the arrangement in figure 85(c). The liquid is contained in a cylinder one end of which is closed by a movable piston. The bulk movement of the liquid is blocked by a membrane, which is impermeable to the liquid, but permeable to the vapour. The whole is maintained at a temperature $T$, so that heat is available for evaporation. If the piston is compressed with a pressure $\pi$ until one mole of liquid has been forced to evaporate, the liquid diminishes in volume by amount $V_L$; the piston does work $\pi V_L$ on the liquid. Consequently the thermal energy necessary for evaporation is reduced from $L$ to $L - \pi V_L$. The vapour pressure then becomes

$$p = A \exp\left[-\frac{(L - \pi V_L)}{RT}\right] = \left[A \exp\left(\frac{-L}{RT}\right)\right] \exp\left(\frac{\pi V_L}{RT}\right) \tag{10.21}$$

$$= p_0\left(1 + \frac{\pi V_L}{RT} + \cdots\right) = p_0 + \frac{p_0 \pi V_L}{RT}.$$

Hence

$$\Delta p = p - p_0 = \frac{p_0 \pi V_L}{RT}. \tag{10.22}$$

If the vapour behaves like a perfect gas $p_0 V_V = RT$ so that we may insert $p_0 = RT/V_V$ in equation (10.22). This gives

$$\Delta p = \pi \frac{V_L}{V_V} = \pi \frac{\rho_V}{\rho_L}. \tag{10.23}$$

Since at room temperature $\rho_V/\rho_L$ is of the order of $10^{-3}$, the increase in vapour pressure with external pressure is generally small. Descriptively one might say that the external pressure squeezes out the liquid molecules. It is not so easy to see why a negative pressure reduces the ease of escape of the liquid molecule. Nevertheless it is clear from the preceding experiment that a negative pressure will in practice reduce the vapour pressure.

## 10.10 Surface tension

### 10.10.1 *Surface energy*

It is common experience that liquids tend to draw up into drops; small drops form perfect spheres. Since a sphere is the geometric form which has the smallest surface area for a given volume we conclude that the surface of the liquid has a higher energy than the bulk; the liquid is therefore always attempting to attain its lowest energy state by reducing its surface area. This energy excess is called the free surface energy $\gamma$ and is measured in J m$^{-2}$. For most simple liquids $\gamma$ has a value of the order of 30–100 mJ m$^{-2}$. This unit is identical with the older unit, still current amongst physical chemists, of erg cm$^{-2}$ or dyne cm$^{-1}$.

If a liquid has area $A$ its free surface energy is $\gamma A$. If we increase the surface by doing work on it the work done is

$$\frac{d(\gamma A)}{dA} = \gamma + A\frac{\partial \gamma}{\partial A}. \tag{10.24}$$

With a liquid, however much we stretch the surface, the surface structure regains its initial equilibrium configuration in a very short interval of time. Thus the equilibrium structure remains unchanged so that $\partial \gamma / \partial A = 0$. This is in contrast with a solid.

Hence

$$\text{work done per unit area of extension} = \gamma. \tag{10.25}$$

Recent experiments have shown that very thin films of water have the same surface energy as two surfaces separated by bulk water. The identity holds down to a thickness of only 2 nm. This implies that the molecular forces responsible for the surface energy (see below) are very short range.

The free surface energy is equivalent to a line tension acting in all directions parallel to the surface. Consider a liquid surface (as shown in figure 86) of width $L$ and length $X$. Suppose at the barrier AB we apply a force $F$ parallel to the surface and normal to AB so that we extend the length of the surface by an amount $x$. The area increase is $Lx$ and the work done in increasing the surface area is $Lx\gamma$. This must be equal to the external work done $Fx$.

Hence

$$Fx = Lx\gamma,$$

Figure 86. Sketch showing equivalence of surface energy per unit area and the tension per unit length.

or

$$\frac{F}{L} = \gamma. \tag{10.26}$$

Thus the surface energy is equivalent to a surface tension or line tension $\gamma \, \text{N m}^{-1}$.

### 10.10.2 *Molecular interpretation*

The free surface energy of a liquid lends itself to a very simple molecular interpretation. Molecules in the bulk are subjected to attraction by surrounding molecules; the field is symmetrical and has no net effect. At the surface, however, the surface molecules are pulled in towards the bulk of the liquid. Apart from a few vapour molecules there is no attraction in the opposite direction. Consequently if we wish to increase the area we have to pull molecules up to the surface from the bulk against this one-sided attraction. This accounts for the surface energy.

Alternatively we may say that in moving molecules from the bulk towards the surface we are continuously breaking and reforming bonds. At the top, however, roughly half the bonds are not reformed. Since bond formation implies a decrease in energy the incomplete reformation of bonds at the surface means that the surface has a higher energy than the bulk. Practically the whole of this process is confined to the last few molecular layers and it is for this reason that extremely thin liquid films have the same surface energy as 'bulk' liquid (see above). Without going into further details it is evident that the surface energy is of the order of one-half the energy required to break all bonds, i.e. one-half the latent heat of vaporization of a molecular layer. This is essentially the same as the model we discussed earlier for solids and we emphasize again that the intermolecular forces responsible for the surface tension of a liquid are of the same nature as those responsible for the pressure defect of the corresponding gas.

If $L$ is the latent heat of vaporization per mole and $M$ the molecular weight and $\rho$ the density, we have, as an approximate relation (cf. equation (6.9)),

$$\gamma = 0.3 \frac{L}{N_A} \left( \frac{N_A \rho}{M} \right)^{\frac{2}{3}}. \tag{10.27}$$

For liquid argon this gives a value of 14 mJ m$^{-2}$ for the surface tension compared with the observed value of 13; for liquid neon 4 compared with 5.5; for nitrogen 11 compared with 10.5; for oxygen 13 compared with

18; for benzene 110 compared with 40 and for mercury 630 compared with 600 mJ m$^{-2}$.

Thermodynamics provides another quantity, the total surface energy $h$. When a surface is increased in area, apart from the surface energy term $\gamma$, there is also a 'latent heat' term. Heat must be provided to keep the temperature constant. Its magnitude is $-T\partial\gamma/\partial T$. Hence the total surface energy is

$$h = \gamma - T\frac{\partial \gamma}{\partial T}.\tag{10.28}$$

Equation (10.27) does not distinguish between these two concepts; it is really a model for $\gamma$ at 0 K so that the poor agreement with the room temperature experimental values must not be taken too seriously. On the other hand the difference between $h$ and $\gamma$ is often significant. For example for water $\gamma = 72$ and $h = 118$ mJ m$^{-2}$. If fine droplets of water are allowed to coalesce so as to destroy their surface area the temperature rise is determined by $h$, not by $\gamma$. This provides a very ingenious method of determining areas of fine particles. They are first equilibrated with water vapour so that their surface is fully covered with a condensed film of water only a few angstroms thick. They are then immersed in water in a delicate calorimeter. If the heat given out is $\Delta Q$ J the surface area is given by $\Delta Q/0.118$ m$^2$.

### 10.10.3 *Contact angle and contact equilibrium*

It is generally found that liquids with low surface tension readily wet most solids, giving a contact angle of zero; those with high surface tension often show a finite contact angle. In molecular terms we may say that if the cohesion between the molecules of the liquid is greater than the adhesion between the liquid and solid, the liquid will not 'want' to wet the solid, consequently it will show a finite contact angle.

Contact equilibrium is shown in figure 87. The free surface of the liquid has a surface energy $\gamma_L$; the solid–liquid interface a surface energy $\gamma_{SL}$

Figure 87. (*a*) Contact angle equilibrium. (*b*) Contact on a thick specimen. (*c*) Contact on a thin mica sheet where the vertical component of the tension distorts the shape of the sheet.

and the exposed portion of the solid adjacent to the liquid where vapour has been adsorbed a surface energy $\gamma_{SV}$. If we consider a virtual movement of the contact region so that an additional $m^2$ of solid is wetted we have

surface energy increase at solid–liquid interface = $\gamma_{SL}$,

surface energy decrease at solid–vapour interface = $\gamma_{SV}$,

surface energy increase of water surface = $\gamma_L \cos \theta$.

By the principle of virtual work

$$\gamma_{SV} = \gamma_{SL} + \gamma_L \cos \theta. \qquad (10.29)$$

This equation, which was first derived by T. Young, without proof, in 1805 and by A. Dupré in 1869, is exactly what we would obtain if we took the horizontal components of the surface tension forces. This at once raises the question: what has happened to the vertical component $\gamma_L \sin \theta$? The answer is simple. The force is very small and its effect on a solid is negligible. If, however, the solid consists of a very thin flexible sheet, for example, of mica, the vertical forces may be sufficient to distort the surface visibly. One is then aware of the reality of the vertical component.

Equation (10.29) has been studied experimentally by Miss S. Kay, using the bifurcation of mica as a means of determining directly the surface energies of the solid. In the first experiment she determined $\gamma$ for a mica surface exposed to water vapour; this gives $\gamma_{SV}$. In the second experiment she determined $\gamma$ when the surfaces were immersed in water; this gives $\gamma_{SL}$. Since water completely wets mica ($\theta = 0$) we should expect to find, from equation (10.29) that

$$\gamma_L = \gamma_{SV} - \gamma_{SL}. \qquad (10.30)$$

Results obtained with water are given in Table 29. The table includes results for hexane which also wets mica. The agreement is very satisfactory.

Table 29. *Surface energy of mica in presence of vapour and liquid*

| Fluid | Surface energies (mJ m$^{-2}$) | | | |
| --- | --- | --- | --- | --- |
| | $\gamma_{SV}$ | $\gamma_{SL}$ | $\gamma_{SV} - \gamma_{SL}$ | $\gamma_L$ |
| Water | 183 | 107 | 76 | 73 |
| Hexane | 271 | 255 | 16 | 18 |

### 10.10.4 *Pressure difference across a liquid-vapour interface*

We shall now discuss a few of the general properties of capillarity which are shown by liquids. Before doing so we derive a simple relation for the pressure difference across a curved liquid interface.

Let ABCD be a portion of a surface marking the boundary between air below and liquid above (see figure 88). Over a small region a curved surface can always be described in terms of two radii of curvature in mutually orthogonal planes. These are known as the principal radii of curvature $R_1$ and $R_2$. Suppose we assume that the pressure on the air side is $p_1$ and on the liquid side is $p_2$. We now apply the principle of virtual work displacing the surface parallel to itself through a distance $dx$. Then AB increases in length by $dx\,\theta_1 = dx(AB/R_1)$, similarly AD increases in length by $dx\,\theta_2 = dx(AD/R_2)$. The increase in surface area is then

$$\Delta\alpha = \left(AB + AB\frac{dx}{R_1}\right)\left(AD + AD\frac{dx}{R_2}\right) - AB \times AD$$

$$\Delta\alpha = AB \times AD \times dx \times \left(\frac{1}{R_1} + \frac{1}{R_2}\right). \tag{10.31}$$

This involves an increase in surface energy of $\gamma\Delta\alpha$. The work done by the pressure difference $p_1 - p_2$ is

$$(p_1 - p_2)AB \times AD \times dx = \gamma\,\Delta\alpha.$$

Figure 88. Pressure difference across a liquid–air interface.

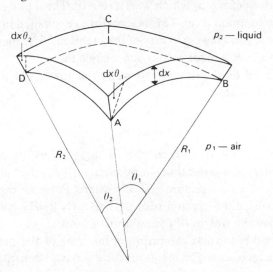

Therefore

$$p_1 - p_2 = \gamma\left(\frac{1}{R_1} + \frac{1}{R_2}\right). \tag{10.32}$$

The pressure on the concave side is greater than on the convex side of a liquid interface, by an amount depending on the surface tension and on the curvature. This relation becomes clearer in the following two very simple cases. For a spherical bubble of radius $R$ within a liquid the surface area is $4\pi R^2$. If a small expansion occurs increasing $R$ by $dR$ the increase in area of surface is $8\pi R\ dR$. The work that must be done by the pressure in the bubble is $4\pi R^2 p\ dR$. Equating this to the increase in surface energy we have

$$4\pi R^2 p\ dR = 8\pi R\ dR\gamma,$$

$$p = \frac{2\gamma}{R}. \tag{10.33}$$

This is exactly the result of equation (10.32) if $R_1 = R_2$. The second case is for a soap bubble. Here two surfaces are involved and the pressure inside the bubble is

$$p = \frac{4\gamma_s}{R}, \tag{10.34}$$

where $\gamma_s$ is the surface tension of the soap solution.

### 10.10.5 *Capillary rise*

Suppose a uniform tube of small radius $R$ is held vertically and lowered into a liquid of density $\rho_L$ which wets it ($\theta = 0$). The liquid rises to a height $h$ above the common level. This is sometimes explained in terms of the surface forces around the periphery of the meniscus. The meniscus has peripheral length $2\pi R$ and for zero contact angle the upward force is $2\pi R\gamma$. This is balanced by the downward weight of the liquid column $h\pi R^2 \rho_L g$ (figure 89(a)). Thus

$$h\rho_L g = \frac{2\gamma}{R}. \tag{10.35}$$

Although equation (10.35) is correct, this is largely fortuitous. For example if the capillary was conical in shape (figure 89(b)) the capillary rise would still be given by equation (10.35) where $R$ is the capillary radius at the point where the meniscus formed. Clearly $2\pi R\gamma$ could be considerably less than the weight of the column of liquid.

The proper method is to treat the problem in terms of the pressure difference across the meniscus. The meniscus for $\theta = 0$ is a hemisphere of radius $R_1 = R_2 = R$. According to equation (10.32) the pressure at B will be less than at A (see figure 89($c$)) by the amount

$$\gamma\left(\frac{1}{R_1} + \frac{1}{R_2}\right) = \frac{2\gamma}{R}. \tag{10.36}$$

But the pressure at A is the same as that at A' (it is atmospheric) and is the same as that at C except for a minute air pressure drop A'C. Thus the pressure in the liquid at B is less than the pressure in the liquid at C. Consequently the liquid is drawn up the tube to a height $h$, where

$$h\rho_L g = \frac{2\gamma}{R}.$$

We should obtain the same result for a conical capillary in agreement with experimental observation.

This treatment also explains why the meniscus in a narrow capillary can be considered to be spherical in shape. If the capillary rise is large compared with the diameter of the tube, the height of rise is practically constant over the whole width of the capillary, i.e. the pressure defect $h\rho_L g$ is constant. This means (see equation (10.36)) that the curvature of the meniscus must be constant. For a circular capillary the two principal radii of curvature must, by symmetry, be equal. Consequently the meniscus will be a portion of a sphere of almost constant radius, the value of which depends on the contact angle. For zero contact angle the shape will be hemispherical, the radius of curvature being equal to that of the tube

Figure 89. ($a$) Simple calculation of capillary rise in terms of surface tension forces. ($b$) Inapplicability of method ($a$) to a conical capillary. ($c$) Calculation of capillary rise in terms of pressure defect.

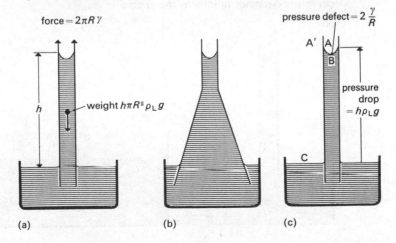

$R$. If the contact angle is $\theta$, it will be equal to $R/\cos\theta$ (see figure 90($a$)) and the pressure defect giving rise to the capillary ascent will be equal to $2\gamma\cos\theta/R$. As the capillary gets wider the height of the liquid column becomes small compared with the capillary diameter (see figure 90($b$)). There is a very large difference between the pressure defect at L and at M. Indeed at M where the height of rise is negligible the curvature is almost zero ($R\to\infty$) whilst it is finite at L and N. The detailed shape of the meniscus can be solved theoretically by equating the pressure defect $\gamma(1/R_1+1/R_2)$ at any given point to the value of $h\rho_L g$ at that point, where $h$ is the height of the liquid above the bulk level of the liquid at a large distance from the walls of the container.

### 10.10.6 *Vapour pressure over a liquid meniscus*

We see from figure 89($c$) that the vapour pressure over the liquid at A is less than over the liquid at C by $h\rho_V g$ where $\rho_V$ is the mean density of the vapour. We may write:

vapour pressure defect

$$= h\rho_V g = h\rho_L\times\frac{\rho_V}{\rho_L}\times g = \frac{2\gamma}{R}\times\frac{\rho_V}{\rho_L}. \tag{10.37}$$

This result is part of a more general statement: the vapour pressure over a concave meniscus surface is less than that over a flat liquid surface by the quantity given in equation (10.37). Similarly the vapour pressure over a convex meniscus is greater than over a flat surface by a similar amount.

Figure 90. ($a$) Meniscus shape for large capillary rise must be of almost constant curvature. ($b$) For small capillary rise the radius of curvature becomes infinite at the centre.

We may see this in a different way by considering the behaviour of a spherical liquid drop of radius $R$. The liquid within the drop is subjected to a surface tension pressure of amount $p = 2\gamma/R$. As we saw in a previous section this increases the vapour pressure by the amount

$$p\frac{\rho_V}{\rho_L} = \frac{2\gamma}{R} \times \frac{\rho_V}{\rho_L}. \tag{10.38}$$

## 10.11 Nucleation in condensation: the Wilson cloud chamber

Consider the equilibrium between a vapour and its liquid and the problem of condensation. If there are already present a few droplets of radius of curvature $r$, the droplets exert a vapour pressure of $(2\gamma/r) \times (\rho_V/\rho_L)$. This can be very large. For example with water vapour at 293 K, $\rho_V/\rho_L$ is about $2 \times 10^{-5}$. But if $r$ is only 0.5 nm the factor $2\gamma/r$ has a value of about $3 \times 10^8$ N m$^{-2}$ so that the droplets are attempting to evaporate with a vapour pressure of about $6 \times 10^3$ N m$^{-2}$. Since the saturated vapour pressure of water at 293 K is about $2 \times 10^3$ N m$^{-2}$, the droplets will evaporate; they cannot grow by further condensation. There are only two ways by which condensation can be made to occur. One is to provide artificial nuclei with relatively large radii of curvature: this is, in practice, generally achieved by dust particles. The second is to produce super-saturation of the vapour, usually by sudden cooling. This is the basis of the Wilson cloud chamber.

Consider a vessel closed with a movable piston in which the vapour is in equilibrium with the liquid. Charged particles are projected through the vapour. Because the electrostatic energy of a point charge is lowered by the condensation of a dielectric on it, some condensation may occur. If the piston is now suddenly raised there is a marked adiabatic temperature drop. Although there is some volume increase the main effect is that the existing vapour pressure is much higher than the saturated vapour pressure at the new reduced temperature. Condensation occurs readily and the path of the particles is revealed as a series of liquid droplets.

## 10.12 Superheating

A similar problem is involved in the heating of a liquid to produce boiling. If the initial bubbles have a radius of curvature $r$, the vapour pressure in the bubbles must exceed $2\gamma/r$ if the bubbles are to grow. For a bubble of radius $r = 0.5$ nm in a liquid such as water, this pressure is of the order of 3000 atmospheres. Thus unless the vapour pressure reaches this value the bubble will collapse. To obtain vapour pressures of this order the liquid must be very strongly superheated. With water, temperatures of the order of several 100 K above the normal boiling point are required. Experiments with water from which dissolved gases have been eliminated do, in fact, show that bubbles will not form until temperatures of this order are reached. Of course, once the bubble begins to form at this high vapour pressure, the excess pressure necessary to produce bubble growth begins to decrease as $r$ increases, so that the bubble grows at a catastrophic rate – the liquid explodes.

Steady boiling at the true boiling temperature is achieved by providing suitable nuclei. These are often air pockets trapped at solid surfaces or gases dissolved in the liquid itself.

### 10.13 The energy for capillary rise

We raise here two points concerning capillary rise that are not often discussed in elementary textbooks. The first is, what is the source of the energy for capillary rise? Consider a tube of radius $R$. The liquid rises to a height $h$ given by

$$h = \frac{2\gamma_L}{R\rho g}. \tag{10.39}$$

The potential energy $\varepsilon_1$ gained by the liquid is equivalent to the raising of a mass of liquid $\pi R^2 h\rho$ through a height $\frac{1}{2}h$. Hence

$$\varepsilon_1 = \frac{1}{2}\pi R^2 h^2 \rho. \tag{10.40}$$

This energy comes from the wetting of the walls of the tube by the liquid. Descriptively we may see this in the following way. The surface energy of a solid arises from the asymmetric forces at the free surface. If we cover the surface with another material, for example, a liquid, we reduce this asymmetry and hence reduce the surface energy. The energy liberated in this process is available for pulling the liquid up the tube. We may make this quantitative as follows.

Consider a liquid with contact angle $\theta = 0$. The equilibrium condition, as we saw earlier, is

$$\gamma_{SV} = \gamma_L + \gamma_{SL}. \tag{10.41}$$

If the liquid advances along the tube so as to cover 1 m$^2$ of the surface we destroy 1 m$^2$ of $\gamma_{SV}$ and gain 1 m$^2$ of $\gamma_{SL}$. There is no change in the area of the liquid meniscus. The energy given up by the system is thus

$$\gamma_{SV} - \gamma_{SL}. \tag{10.42}$$

From equation (10.41) we see that this is equal to $\gamma_L$. Thus each m$^2$ of capillary surface wetted releases energy $\gamma_L$. The energy released for a rise of $h$ is then

$$\varepsilon_2 = \gamma_L 2\pi R h. \tag{10.43}$$

Substituting from equation (10.39) for $\gamma_L$ wc have

$$\text{loss of surface energy } \varepsilon_2 = \pi R^2 h^2 \rho g. \tag{10.44}$$

This is exactly twice the gain in potential energy given by equation (10.40). This implies that if the liquid were non-viscous it would rise to a height $2h$ and oscillate between 0 and $2h$ with its mean position at $h$. In practice viscosity consumes the excess energy very quickly.

The behaviour is like that of a vertical spiral spring onto the end of which a weight is suddenly attached. The weight falls instantaneously to double its equilibrium depth and oscillates between this and zero until the kinetic energy is consumed by friction and static equilibrium is achieved.

The second query is the following. Some solids are known to have much higher surface energies than others. For example if they are first thoroughly cleaned and then immersed in a liquid the amount of heat released may vary very widely; nevertheless if a liquid wets these surfaces ($\theta = 0$), the capillary forces do not depend on the material of which the capillary is made; they depend only on the surface energy of the liquid. Why is this? The answer is implicit in equation (10.41). The adsorbed vapour on the solid will always adjust its thickness so that the quantity

$$\gamma_{SV} - \gamma_{SL}$$

is equal to $\gamma_L$ (for $\theta = 0$). The energy gain is thus dependent only on $\gamma_L$. We may describe this differently. The vapour adsorbed on the solid presents a surface to the advancing liquid which closely resembles the liquid itself. On a high-energy surface the adsorbed film will tend to be thicker than on a low-energy surface. In both cases (for $\theta = 0$) the liquid is virtually wetting itself so that only $\gamma_L$ is involved in the process. In this connection we may note that the range of forces is small and that the adsorbed film needs to be only 1–2 nm thick in order to mask almost completely the influence of the underlying substrate.

# 11

---

## *Liquids: their flow properties*

In this chapter we shall discuss the flow properties of liquids, and derive or state some of the more standard equations of flow. However, our main attention will be directed to a molecular model of viscous flow; this follows closely the theory proposed by Eyring.

### 11.1  Flow in ideal liquids: Bernoulli's equation

As with gases we can assume the existence of ideal liquids in which internal forces play a trivial part, so that they have negligible surface tension or viscosity. Their flow properties are determined solely by their density. Of course, no liquids can have zero internal forces, they would not be liquid if this were so, but they can often behave as ideal liquids in flow if the inertial forces dominate.

Consider the flow of such an idealized liquid and let us follow the path of any particle in it. If it moves in a continuous steady state we may draw a line such that the tangent at any point gives the direction of flow of the particle. Such lines are called streamlines. They are smooth continuous lines througout the liquid and can never intersect – no fluid particles can flow across from one streamline to another. Let AB represent an imaginary tube in the liquid bounded by streamlines (figure 91($a$)). At A the liquid is at a height $h_1$ above a reference level, the pressure acting at that point is $p_1$, the cross-sectional area of the tube is $\alpha_1$, and the liquid flow velocity is $v_1$; at B the corresponding quantities are $h_2$, $p_2$, $\alpha_2$, $v_2$. Consider the energy balance during a short time interval d$t$. The pressure $p_1$ at A drives a volume of liquid $\alpha_1 v_1$ d$t$ through the tube: the work done is

$$w_1 = p_1 \alpha_1 v_1 \, \mathrm{d}t. \tag{11.1}$$

The mass of liquid involved is $\rho \alpha_1 v_1 \, dt$ so that the kinetic energy it brings with it is

$$w_2 = \tfrac{1}{2}(\rho \alpha_1 v_1 \, dt)v_1^2,$$  (11.2)

and its potential energy is

$$w_3 = (\rho \alpha_1 v_1 \, dt)gh_1.$$  (11.3)

Since the motion is assumed steady and the liquid is ideal the same amount of energy leaves the tube at B. Hence

$$(w_1 + w_2 + w_3)_A = (w_1 + w_2 + w_3)_B.$$  (11.4)

Because the same mass of liquid leaves at B as enters at A,

$$\alpha_1 v_1 \, dt = \alpha_2 v_2 \, dt.$$  (11.5)

Equation (11.4) becomes

$$\frac{p_1}{\rho} + \tfrac{1}{2}v_1^2 + h_1 g = \frac{p_2}{\rho} + \tfrac{1}{2}v_2^2 + h_2 g.$$  (11.6)

This is Bernoulli's equation of flow for an ideal liquid: the three terms on each side of the equation correspond respectively to the pressure energy to move the liquid, the kinetic energy and the potential energy.

A simple example is illustrated in figure 91(b). A container is filled with liquid to a height $h_1$. At the bottom of the container is a small hole. What is the velocity $v_2$ of the outflowing liquid? If we construct a tube of flow as shown, the conditions at the top of the tube are $p_1 =$ atmospheric $= A$,

Figure 91. (a) Streamlines of flow of a liquid in a gravitational field. (b) Flow of liquid from an orifice at a depth $h$ below the free level ($p_1 = p_2 =$ atmospheric pressure $A$).

height $= h_1$, velocity, $v_1 \approx 0$. At the bottom of the tube $p_2 = A$, height $= 0$, velocity $= v_2$. Starting with equation (11.6),

$$\frac{p_1}{\rho} + \tfrac{1}{2}v_1^2 + h_1 g = \frac{p_2}{\rho} + \tfrac{1}{2}v_2^2 + h_2 g,$$

$$\frac{A}{\rho} + 0 + h_1 g = \frac{A}{\rho} + \tfrac{1}{2}v_2^2 + 0,$$

$$v_2^2 = 2g h_1. \tag{11.7}$$

The flow of an ideal liquid does not involve any particular molecular model; density is the only property concerned, and we shall not discuss it further. We may, however, remark that in general an increase in flow velocity is accompanied by a drop in pressure. This accounts for the suction experienced by a ship approaching a quay – the water flows faster at the regions of nearest approach. It also accounts for the downward curvature of the flight of a tennis ball that has been given top spin. Again, because the top surface of an aerofoil is longer, the air velocity in flight is higher and the pressure lower than over the bottom surface; this produces aerodynamic lift.

## 11.2 Flow in real liquids, viscosity
### 11.2.1 Steady and unsteady flow, turbulence

Real liquids experience a viscous resistance to flow. We have already defined viscosity in discussing the behaviour of gases but we may recapitulate it here. If the velocity gradient between two neighbouring planes is $du/dz$ the force per $m^2$ to overcome viscous resistance is

$$f = \eta \frac{du}{dz}, \tag{11.8}$$

where $\eta$ is defined as the viscosity. Fluids which obey this relationship are known as Newtonian fluids. The dimensions of $\eta$ are

$$[\eta] = ML^{-1}T^{-1}. \tag{11.9}$$

If the flow is steady and stable the work done is expended solely in overcoming the viscosity, and the work appears as heat. If, however, the velocity of flow is too high or there are other unfavourable features, a series of vortices may develop in the liquid. Some of the work is expended in providing kinetic energy for the vortices. Vortices often assemble on a solid boundary and this is known as the boundary layer. The condition for turbulent motion was first established by O. Reynolds. If a liquid of density $\rho$, viscosity $\eta$ flows along a channel of lateral dimension $a$, there is a critical velocity at which orderly streamlined flow changes to turbulent motion. By dimensional analysis it is easy to show that this occurs when

$$v_c = \frac{k\eta}{\rho a}. \tag{11.10}$$

The quantity $k$ is called Reynolds' number. It is dimensionless and the transition from streamline to turbulent motion occurs when its value is around 1000 to 2000.

For $\quad \frac{v\rho a}{\eta} < k \quad$ the flow is streamlined.

For $\quad \frac{v\rho a}{\eta} > k \quad$ the flow is turbulent.

Turbulence is essentially a *condition* of instability and the force involved in flow is not necessarily a clear indication of whether turbulence has begun or not. The force is determined primarily by viscous or inertial factors. An interesting simple example of these two conditions is furnished by considering the steady movement of a solid sphere of radius $a$ through a fluid. The resistance to motion can be derived by dimensional arguments. One solution for the resistive force is

$$F = Aa\eta v. \tag{11.11}$$

This is Stokes' law; analysis shows that $A$ has the value $6\pi$.
Another solution is

$$F = Bv^2\rho a^2. \tag{11.12}$$

This does not involve viscosity but kinetic energy, the force being determined by the momentum of the oncoming fluid.

### 11.2.2 *Viscous flow through a tube*

Consider a cylindrical tube of length $l$, radius $r$. Liquid enters one end at pressure $p_1$ and leaves the other end at pressure $p_2$. If the flow is uniform the streamlines are all parallel to the tube axis, there is no slip of liquid at the solid boundary and the velocity profile is parabolic. The rate of flow of liquid per second is

$$Q = \frac{\pi r^4}{8\eta}\left(\frac{p_2 - p_1}{l}\right). \tag{11.13}$$

Poiseuille first used this equation to determine the viscosity of blood in horses' arteries.

### 11.2.3 *Rotation of a shaft in a bearing: hydrodynamic lubrication*

Consider a shaft of radius $a$ rotating with angular velocity $\omega$ in a bearing of radius $a + c$ (figure 92($a$)). If the shaft carries a light load $W$ and the space between shaft and bearing is filled with oil, they will remain concentric. The velocity gradient across the film is $\omega a/c$. The viscous force $f$ per m$^2$ is this quantity multiplied by $\eta$. If the length of the shaft is $l$, the resultant tangential force that must be exerted at the edge of the shaft to maintain steady rotation is

$$F = f \times 2\pi al = \frac{\omega a}{c} \eta 2\pi al = \frac{Na}{c} \eta al \times 4\pi^2, \tag{11.14}$$

where $N = \omega/2\pi$ is the number of revolutions per second.

We define the coefficient of friction $\mu$ as the ratio of $F$ to $W$. Then

$$\mu = \frac{Na}{c} \eta \frac{al}{W} \times 4\pi^2. \tag{11.15}$$

Since the nominal pressure $P$ on the bearing is $W/2al$ we may write

$$\mu = \frac{4\pi^2 N\eta}{P} \left(\frac{a}{2c}\right). \tag{11.16}$$

For a bearing and shaft of fixed dimensions the quantity $a/2c$ is a constant so that $\mu$ is linearly proportional to $N\eta/P$ (see figure 92($b$)).

This relation was first derived by A. A. Petroff in 1883 and describes adequately the behaviour of a lightly loaded bearing. However, it is evident that for loads which are not negligible, the journal cannot run in the central position. If the oil film is to support an appreciable load the journal must rotate eccentrically relative to the bearing, so that the lubricant is squeezed through the converging gap between the surfaces. This convergence in a viscous liquid produces a hydrodynamic pressure in the oil wedge as O. Reynolds showed in 1886 and under suitable conditions it is sufficient to support the applied load.[†]

For a journal in a 'half' bearing the pressure distribution is as shown in figure 92($c$) and similar results are obtained for a 'full' bearing. It is seen that the (asymmetric) pressure distribution gives a resultant upthrust which would not occur if the journal rotated centrally in the bearing. Nevertheless the viscous resistance is not very different from that given by

---

† If the liquid were completely non-viscous the rotating journal would not drag any liquid with it. If we were to supply an external pressure and force the liquid through the gap, the convergence, as we saw in section 11.1, would actually produce a pressure *drop* in the convergent gap. It is primarily the viscous property of the liquid that is responsible for the development of a positive pressure in the liquid wedge.

the simple Petroff theory and, for a well-designed bearing, friction coefficients of the order of 0.001 are obtained.

A far more important difference between the Petroff and Reynolds treatment concerns the distance of nearest approach of journal to bearing. As the load increases the oil convergence must increase and this leads to a decrease in the separation. At some critical stage the separation necessary for ideal hydrodynamic lubrication may be less than the height of the surface roughnesses. The hydrodynamic film is penetrated and contact occurs between the metals themselves or, at best, between metals covered by a thin adsorbed lubricant layer only one or two molecules thick. This is referred to as boundary lubrication and involves higher friction ($\mu = 0.1$) and possible wear. By contrast hydrodynamic lubrication gives a very low coefficient of friction ($\mu \sim 0.001$) and, in principle, no wear at all. The region between the classical hydrodynamic and boundary regime is often referred to as the mixed regime. More recent work shows that in this region contact between asperities may be prevented by another mechanism. The very high pressures generated at the

Figure 92. (*a*) Rotation of a shaft of radius *a* in a bearing of radius $a + c$ (Petroff). If the liquid has viscosity $\eta$ and the angular velocity is $\omega$ the viscous force $F$ that has to be overcome is proportional to $\eta\omega$. (*b*) If the normal load is $W$ the coefficient of friction is proportional to $\eta\omega/F$ or to $N\eta/P$, where $N$ is the number of revolutions per second and $P$ the normal pressure. At high pressures or low speeds the hydrodynamic film is penetrated and the friction rises. (*c*) In practice, a convergent oil wedge must be formed so that sufficient pressure can be developed in the liquid to support the applied load; as a result the shaft is not quite concentric with the bearing (Reynolds).

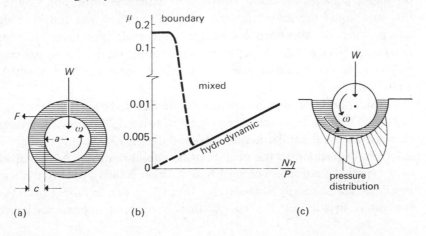

(a)  (b)  (c)

points of imminent contact can produce a prodigious local rise in viscosity. Consequently the lubricant cannot be extruded from precisely those areas where metallic contact might otherwise occur. In this way the range of successful lubrication is extended into a region of far more severe conditions. This type of lubrication, where high elastic stresses have a favourable effect on the viscosity of the lubricant, is known as elasto-hydrodynamic lubrication. When the elasto-hydrodynamic film itself is finally penetrated or destroyed we are back again at classical boundary lubrication where only molecular adsorbed layers remain.

### 11.2.4 *Simple ideas on the viscosity of liquids*

The viscosity of gases is explained in terms of molecular momentum-transfer from one plane to a neighbouring plane. The mechanical analogy is to consider two trains travelling parallel to one another at slightly different speeds. People are continuously jumping from one train to the other, so that those leaving the slower train, slow up the faster; whilst those jumping from the faster train speed up the slower. There is no net change in the numbers of people in any one train but there is clearly a momentum transfer tending to equalize the velocities of the two trains. Work must be done on the trains to maintain their difference in velocity. It is this process on a molecular scale which is responsible for the viscosity of gases.

In liquids the momentum transfer process also takes place but it is a trivial part of the viscous resistance. The train model may still be used but now the jumping passengers find themselves caught by their coat-tails. They must grab hold of the other train and pull very hard before they can detach themselves from their own train. This will have a marked effect in tending to equalize the train velocities. In molecular terms one may say that in a liquid the molecules are so close together that considerable energy must be expended in dragging one molecular layer over its neighbour. No rigorous theory exists but a simple molecular model may be given. We first give it descriptively and in the next section deal with it analytically.

Consider an instantaneous picture of the liquid. There is short-range order and we may represent a small part of two neighbouring molecular planes by an array similar to that of a solid (figure 93). To drag molecule A2 from its position to the neighbouring equilibrium site A2′ involves overcoming the attraction of neighbours. There is thus a potential hill to be climbed, and energy is needed for this. (Note that in falling from the summit X to site A2′ the energy that is restored appears as heat.)

Occasionally molecules will have enough thermal energy to do this but there is just as much chance of the molecule jumping to site A1 as to A2'. There is thus no net flow in any direction. If now we apply a shear stress to the right on the molecules in plane A mechanical work is done on the molecules. The thermal energy necessary to move the molecule to the right is reduced whilst that required to move it to the left is increased (figure 93(b)). Thus thermally activated flow to the right is favoured. Another way of looking at the process is to say that before molecule A2 can move to site A2' it must make available a vacancy in the liquid to receive it. The energy to do this depends on the amount of opening up of the liquid that is necessary and on the latent heat of vaporization. Indeed the energy needed to make a vacancy available is roughly equal to $\frac{1}{3}$–$\frac{1}{2}$ of the latent heat of vaporization of the vacancy.

### 11.2.5 *The Eyring theory of liquid viscosity*
We consider a small cluster of molecules as shown in figure 93. In order to transpose molecule A2 to position A2' against the attractive forces of neighbours we have to surmount a potential energy barrier $\varepsilon$.

Figure 93. The Eyring molecular model of liquid viscosity $(a)$ to transpose a molecule from position A2 to A2' intermolecular forces must be overcome. This involves a potential energy barrier $\varepsilon$, $(b)$ if a shear stress is applied to the right the potential energy curve is distorted, transitions to the right are facilitated, to the left hindered, $(c)$ a shear stress $f$ acting on a molecule occupying a nominal area $\alpha$ exerts a shear force $f\alpha$.

The rate at which this is possible as a result of thermal fluctuations is

$$C \exp\left(-\frac{\varepsilon}{kT}\right),$$
(11.17)

where $C$ is a frequency term. Eyring suggests that this is the vibrational frequency of the molecule and is given approximately by $kT/h$, where $h$ is Planck's constant. Hence the frequency with which a molecule can jump to a neighbouring site as a result of thermal activation is

$$\mathcal{N}_0 = \frac{kT}{h} \exp\frac{-\varepsilon}{kT}.$$
(11.18)

When no stress is applied to the liquid there is an equal rate of jumping to left and right so that no net flow occurs. Suppose now a shear stress $f$ is applied (see figure 93(c)). If the average area occupied by a molecule is $\alpha$ the force on the molecule is $f\alpha$. The only mechanical work we can perform is expended in carrying the molecule to the top of the potential barrier (on the other side of the barrier peak the molecule is assumed to give up all its energy as heat). The work done is

$$\frac{f\alpha\lambda}{2},$$
(11.19)

where $\lambda$ is the distance between neighbouring molecules in the plane.

It follows that the thermal energy required for a jump to the right is reduced from $\varepsilon$ to $\varepsilon - \frac{1}{2}f\alpha\lambda$, i.e. the rate of jumping is increased. Similarly the thermal energy required for a jump to the left is increased to $\varepsilon + \frac{1}{2}f\alpha\lambda$ so that the jumping rate is reduced. The net rate of jumping to the right is now,

$$\mathcal{N} = \frac{kT}{h}\left[\exp\left(-\frac{\varepsilon - \frac{1}{2}f\alpha\lambda}{kT}\right) - \exp\left(-\frac{\varepsilon + \frac{1}{2}f\alpha\lambda}{kT}\right)\right],$$
(11.20)

$$= \frac{kT}{h}\exp\left(-\frac{\varepsilon}{kT}\right)\left[\exp\left(\frac{f\alpha\lambda}{2kT}\right) - \exp\left(-\frac{f\alpha\lambda}{2kT}\right)\right].$$
(11.21)

For the exponentials in the square bracket, if the shear stresses are not too high, i.e. if $f\alpha\lambda \ll kT$, we may take the first terms of the expansion. This gives

$$\mathcal{N} = \frac{kT}{h}\exp\left(-\frac{\varepsilon}{kT}\right)\left(2\frac{f\alpha\lambda}{2kT}\right) = \frac{f\alpha\lambda}{h}\exp\left(-\frac{\varepsilon}{kT}\right).$$
(11.22)

We must now link $\mathcal{N}$ with the macroscopic flow process. The velocity gradient across the two molecular layers separated by a distance $\lambda_1$ is

$$\text{velocity gradient} = \frac{\text{velocity difference}}{\lambda_1}$$

$$= \frac{(\text{distance per jump}) \times (\text{number of jumps per second})}{\lambda_1}$$

$$= \frac{\lambda \mathcal{N}}{\lambda_1}. \tag{11.23}$$

But by definition the viscosity $\eta$ is given by

$$\eta = \frac{\text{force per unit area}}{\text{velocity gradient}} = f \times \frac{\lambda_1}{\lambda \mathcal{N}}. \tag{11.24}$$

Inserting the value of $\mathcal{N}$ from equation (11.22) we obtain

$$\eta = \frac{h\lambda_1}{\alpha\lambda^2} \exp\left(\frac{\varepsilon}{kT}\right). \tag{11.25}$$

Putting $\lambda_1 \approx \lambda$ and treating $\alpha\lambda_1$ as the effective volume occupied by a molecule $(v_m)$, equation (11.25) becomes

$$\eta = \frac{h}{v_m} \exp\left(\frac{\varepsilon}{kT}\right)$$

or

$$\eta = \frac{hN_A}{V_m} \exp\left(\frac{E}{RT}\right), \tag{11.26}$$

where $N_A$ is Avogadro's number, $V_m$ the volume per mole of the liquid and $E$ the molar activation energy for surmounting the energy barrier. This relation is in full accord with the very marked decrease in viscosity observed with increasing temperature.

The barrier energy $E$ may also be regarded as the energy to create a hole in the liquid big enough to receive a molecule. This ought, therefore, to be comparable with the latent heat of vaporization $(L)$; however, because there is already some free volume in the liquid, the work to open up a molecular hole is less than this. For a very large range of liquids

$$E \approx (0.3 \text{ to } 0.4)L.$$

Consequently for a very wide range of liquids we may write

$$\eta = \frac{hN_A}{V_m} \exp\left(\frac{0.4L}{RT}\right). \tag{11.27}$$

However, for liquid metals it is found that $E$ is very much less than $0.4L$, presumably because the metal ions are much smaller than the metal atoms.

### 11.2.6 *Effect of external pressure on viscosity*

To examine the effect of pressure on liquid viscosity it is simplest to regard the energy barrier $E$ as the energy required to open up a hole for the transposed molecule. In the absence of applied pressure we may write equation (11.26) in the form

$$\eta_0 = D \exp\left(\frac{E}{RT}\right). \tag{11.28}$$

If now an external pressure $P$ is applied the work to form a hole is increased by $PV_h$ where $V_h$ is the volume of the hole. The thermal energy for activated flow is then increased from $E$ to $E + PV_h$. The viscosity then becomes

$$\eta = D \exp\left(\frac{E + PV_h}{RT}\right)$$
$$= \eta_0 \exp\left(\frac{PV_h}{RT}\right). \tag{11.29}$$

This type of relation is found to hold over a fairly wide range of pressures. If $\ln \eta$ is plotted against $P$ a straight line of slope $V_h/RT$ is obtained. It turns out that $V_h$ is about 15 per cent $V_m$ for simple liquids and about 5 per cent $V_m$ for liquid metals. Attempts to explain away these discrepancies are not convincing. One must accept this as part of the inadequacy of the model. In broad outline it is surprisingly successful.

The effect of high pressures in increasing viscosity can be very significant in lubrication. It is probable that many sliding mechanisms operate successfully because local high pressures, which would normally be expected to squeeze out the lubricant and cause seizure, so increase the viscosity of the oil that it remains trapped between the surfaces. For example with heavily loaded gear teeth where elastic deformation in the contact zone may produce pressures of the order of 10 000 atmospheres or $100 \text{ kg mm}^{-2}$, the viscosity of a typical lubricant may be increased a million-fold. In this way nature is much kinder to the engineer than he might have supposed. This type of elasto-hydrodynamic lubrication, (mentioned previously on p. 268) is an area of great importance in current lubrication practice.

### 11.2.7 *Summary*

Although the model of liquid viscosity which we have discussed above is greatly simplified it brings out three important points. Liquid viscosity arises from intermolecular forces and the ability of thermal fluctuations aided by an applied shear stress to produce flow in the direction of shear. Consequently the viscosity decreases if the temperature is increased. Secondly the viscosity increases if the pressure on the liquid is increased. This is because the molecules are pushed closer together, or, in terms of the alternative way of describing the molecular jumps, more work must be done to open up a vacant site.

A third conclusion is that viscosity, being determined by intermolecular forces, is closely related to surface tension $\gamma$ and both are related to the latent heat of vaporization. For many simple organic liquids there is a fair correlation between $\ln \eta$ and $\gamma/Z$, where $Z$ is a factor allowing for the size of the molecule.

It may be mentioned that these models are most suitable for unassociated liquids with van der Waals bonding between the molecules. Highly polar molecules and long complicated molecules do not fit in so well.

The basic differences between the mechanisms responsible for viscosity in gases and in liquids are clearly brought out in Table 30.

### 11.3 **Rigidity of liquids**

The molecules in a liquid jump about with a frequency of $10^{10}$–$10^{12}\,\mathrm{s}^{-1}$. If the rate of shear is sufficiently great there may not be time for the molecules to advance to a neighbouring site. In this case the liquid will not show viscous flow, but will show a finite elastic rigidity. For simple liquids the rates of shear for this to occur are enormous but for liquids with large bulky molecules it may occur at moderate rates of shear. There is a similar behaviour in polymers such as polyethylene, nylon and

Table 30. *Viscosity of gases and liquids*

| Fluid | Effect of temperature $T$ | Effect of pressure $P$ |
|---|---|---|
| Gases | $\eta$ increases as $T^{\frac{1}{2}}$ | None (but see Fig. 20) |
| Liquids | $\eta$ decreases as $$\ln \eta = A + \frac{B}{T}$$ | $\eta$ increases as $$\ln \eta = A + kP$$ |

perspex. The flow of polymeric chains which occurs at slow rates of loading resembles the viscous flow of liquids; this may be impossible at high rates of strain. The material will then deform elastically. A very striking example of this type of behaviour is silicone putty. This material will bounce with a very high resilience, since there is not enough time for the molecules to flow. Again if the material is in the form of a rod and is pulled rapidly it will first stretch and then snap in a brittle fashion. On the other hand at very low rates of deformation, it will flow in a viscous manner.

## 11.4  Non-Newtonian flow

Many liquids encountered in daily life are non-Newtonian. This is particularly true of liquids with large bulky molecules or those which have some additional structure, for example colloids. A finite stress is often needed to break down the structure and initiate flow (see figure 94($a$)). Such materials are known as Bingham fluids. In many cases, once flow starts the molecules tend to orient along the direction of shear. The interaction of the molecules is reduced and the viscosity diminishes as the rate of shear increases (see figure 94($b$)). This is known as thixotropy and often occurs with paints. Brushing reduces the viscosity and the paint flows easily. On the other hand, since the effect is only slowly reversible, the paint will not drip easily when left undisturbed.

Figure 94. Schematic diagram of stress against shear rate for ($a$) a Bingham fluid, ($b$) a Bingham fluid showing thixotropy.

## 11.5 The colloidal state of matter

Before leaving the subject of liquids we consider here briefly the nature of the colloidal state. The colloidal state is very wide spread in biological systems and in many practical situations. Basically a colloid consists of 2 distinct phases, a continuous phase (the dispersion medium) and a particulate phase, where the particles generally have dimensions ranging between 20 and 2000 Å (2 and 200 nm). The two phases can be liquid-in-liquid (milk), solid-in-liquid (paint), liquid-in-gas (aerosol) and other combinations. The basic problem with colloidal systems is their stability. Clearly the particles must not be too large otherwise gravity will produce ready sedimentation. However, if the density of the two phases is similar this tendency will be reduced. The other important factor is the attractive force between the particles: if this is too large the particles will cling together and separate as a cluster from the dispersion. To prevent this various techniques are employed to reduce the attractive forces or to introduce repulsive forces.

We consider here the solid-in-liquid type of colloid. There are two broad classes, lyophilic and lyophobic. The lyophilic (liquid-loving) colloids are those in which there is a strong affinity between the liquid and the particle. The liquid is strongly adsorbed to the surface of the particle so that the interface between the liquid-covered particle and the liquid is very similar to the interface between the liquid and itself. This is intrinsically a relatively stable system particularly if some suitable ions are also available for adsorption on the surface of the particles (see below). In general lyophilic colloids are more stable the smaller the particle size.

In lyophobic (liquid-hating) colloids the liquid does not show affinity for the solid particle. If the forces between the particles are due to van der Waals interactions, it is easy to see that there will be a strong tendency for them to stick together whenever they come into contact, i.e. the colloid will be unstable and coagulation will occur. We recall that in Chapter 1 the van der Waals force between a sphere and a flat surface was shown to be equal to $AR/6x^2$, where $x$ is the distance of nearest approach, $R$ is the radius of the sphere and $A$ the appropriate Hamaker constant. For two identical spheres the force is half this. We now consider the potential energy involved in bringing two such spherical particles together from infinity to a separation $x_0$ comparable with an atomic distance, say 0.3 nm: the potential energy $u$ of the spheres is negative as is always the case for attracting bodies and can be written

$$u = -\int_{x_0}^{\infty} \frac{AR}{12x^2}\, \mathrm{d}x = -\frac{AR}{12x_0}.$$

(11.30)

(We have assumed that only the short range unretarded forces are important and that the integration is valid down to atomic separations.) For typical particles in an aqueous medium $A$ usually has a value of the order of $10^{-20}$ J. Thus if $R$ is, say 100 nm, $u$ has the value $3 \times 10^{-19}$ J. But the thermal energy of the particles is of order $kT$. At room temperature (300 K) this has the value $4 \times 10^{-21}$ J. This is far smaller than the van der Waals potential energy $u$ for particles in atomic contact. Thus once the particles touch they will never be able to separate: the thermal energy will be insufficient to overcome the attractive energy. (The situation resembles the behaviour of a gas below its critical temperature.) Consequently the particles will coagulate unless some method is devised for overcoming the attractive forces between them. This may be achieved by adding, to the liquid phase, ions which can adsorb on the surface of the particles and form a diffuse double layer of charge near the surface. The general principle is as follows. Suppose the particle itself has no net surface charge but that positive ions are preferentially adsorbed on to its surface. In the complete absence of thermal motion an equal number of negative ions (counter ions) would absorb onto the positive charge and neutralize it. Because of thermal agitation such a compact double layer does not form. Instead a diffuse double layer is formed as shown in figure 95. There is a

Figure 95. Distribution of positive and negative ions in a liquid adjacent to a solid surface (the electrical double layer).

fairly rapid change in concentration of negative and positive ions as we move away from the surface (see figure 96($a$)). As a result the potential falls off as shown in figure 96($b$), according to a relation of the form $\psi = \psi_0 \exp(-kx)$. At a distance $x = 1/k$ the potential has fallen to $1/e$ of its value at the surface and beyond this the change in potential is small. Thus $1/k$ may be considered as the thickness of the double layer. The higher the concentration of ions the more the 'centre of gravity' of the charge moves towards the surfaces and the smaller the thickness of the double layer. For example for a monovalent electrolyte the thickness is about 10 Å (1 nm) for a $10^{-1}$ M solution and about 100 Å (10 nm) for a $10^{-3}$ M solution. The higher the temperature the larger the thickness.

If two particles approach one another the electrical double layer will produce a repulsive force that falls off approximately exponentially with separation. This repulsion opposes the attractive forces between the particles arising from van der Waals interactions. To assess the overall effect it is more convenient to deal in potential energies. Figure 97 shows typical potential energies for the van der Waals attraction and the double layer repulsion. In this figure the resultant potential shows a primary deep minimum at $M_1$ corresponding approximately to contact between the particles and a secondary shallower minimum at $M_2$. If the particles are initially far apart and approach one another to a separation $M_2$ and if their thermal energy ($kT$) is small compared with the depth of this minimum

Figure 96. ($a$) Concentration of ions in the double layer as a function of distance from the solid surface. ($b$) Potential within the double layer as a function of distance from the solid surface.

the particles will not be able to escape from one another. The particles will be loosely held together at fairly large separations. This type of instability is known as flocculation. The colloid can be restabilized by heating (increasing $kT$) or by changing the ionic strength to increase the repulsive force. Figure 98 summarizes 4 main types of potential energy curve. In curve A the repulsive forces are weak and the curve is not greatly different from the van der Waals curve itself. These particles will attract one another and reach equilibrium at the primary minimum $M_1$.

Figure 97. Potential energy between two particles in a liquid showing the combined effect of van der Waals attractive forces and electrical double layer repulsive forces. Flocculation at $M_2$.

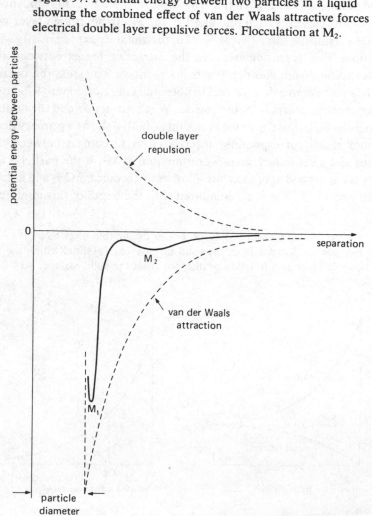

The particles are then virtually in contact and whole clusters will come out of suspension. This is known as coagulation, in contrast to flocculation where the particles are loosely held together at fairly large separations (see figure 97).

In curve B where the repulsive forces have been increased there is a very shallow secondary minimum at $M_2$ but it is so shallow compared with $kT$ that the particles are easily able to escape from one another. There is

Figure 98. Four examples of the resultant potential between two particles. A: immediate contact at $M_1$ (coagulation); B: temporary flocculation at $M_2$ and possibility of coagulation; C: no flocculation or coagulation; D: spontaneous dispersion.

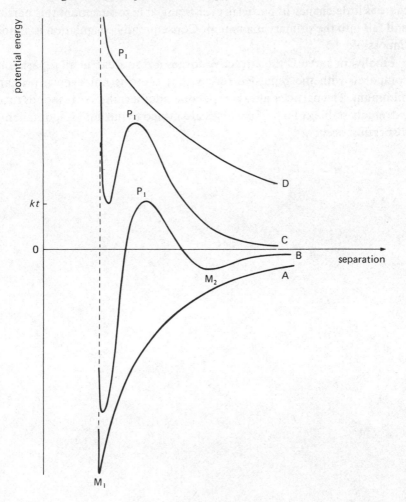

little danger of flocculation. On the other hand the maximum $P_1$ represents an energy barrier comparable with $kT$. Thus particles coming from a large separation may well be able to surmount this barrier and fall to the primary minimum $M_1$. Since this is relatively deep compared with $kT$ the particles will not have enough energy to escape from one another. In that case irreversible coagulation will have occurred. It is interesting to note that paint is a classical example of a flocculated colloid of this type: if left undisturbed for a prolonged period it may become permanently coagulated.

In curve C the repulsive forces are so large that there is no secondary minimum. Consequently there is no tendency even for transient flocculation to occur. Further the maximum $P_1$ is so much greater than $kT$ that there is little chance of particles ever being able to surmount this barrier and fall into the primary maximum. Consequently coagulation is almost impossible.

Finally, in curve D the attractive forces are so weak at all separations compared with the repulsive forces that there is not even a primary minimum. The particles always repel one another: this corresponds to an extremely stable colloid. If particles are added to the liquid, spontaneous dispersion occurs.

# 12

## Dielectric properties of matter

In this chapter we shall describe the main dielectric properties of matter and see how far they can be explained in terms of simple atomic models. As we shall see the main mechanism is the polarization of individual atoms and molecules by an electrostatic field.

### 12.1 Basic dielectric relations
#### 12.1.1 *Dielectric contant (relative permittivity) $\varepsilon$ and polarization $P$*

Suppose we consider a parallel plate condenser carrying a charge $\sigma$ coulombs for each $m^2$ of surface. The field in free space between the plates (see figure 99($a$)) is

$$E_v = \sigma/\varepsilon_0, \tag{12.1}$$

where $E_v$ is measured in $V\,m^{-1}$ and $\varepsilon_0$ is the permittivity of free space and equals $8.85 \times 10^{-12}\,F m^{-1}$. The potential difference between the plates is $\sigma d/\varepsilon_0$ and the capacity $C$ per $m^2$ of condenser is

$$C = \frac{\text{charge}}{\text{potential}} = \frac{\sigma}{\sigma d/\varepsilon_0} = \frac{\varepsilon_0}{d}. \tag{12.2}$$

We now consider what happens if we fill the intervening space with a dielectric. We find that the capacity or relative permittivity is increased by a factor $\varepsilon$. This is known as the dielectric constant of the material.

For an *isolated* condenser, since $\sigma$ remains unchanged, this means that the potential difference between the plates is reduced by a factor $\varepsilon$, so that the field is reduced from $E_v = \sigma/\varepsilon_0$ to

$$E = \frac{\sigma}{\varepsilon_0} \frac{1}{\varepsilon}. \tag{12.3}$$

This reduction in field is due to the polarization of the material. The field distorts the atoms and molecules, and orients existing dipoles in such a way as to create a field opposing the applied field. The resultant field is thus reduced.† We define a polarization $P$ as the electrostatic moment per unit volume induced in the dielectric. As we shall see later $P$ is very nearly proportional to the resultant field $F$. Consider a bar of dielectric of section $A$, length $l$. Its total moment due to polarization is

$$P \times \text{volume} = PAl = PA \times l = \text{end charge} \times \text{length}.$$

Thus the moment per unit volume is equivalent to a surface charge $P$ per unit area. We now see that the effective charge density on the plates is $\sigma - P$ (see figure 99($b$)). The resultant field is then

$$E = \frac{1}{\varepsilon_0}(\sigma - P)$$

or

$$\sigma - P = \varepsilon_0 E. \tag{12.4}$$

Figure 99. ($a$) The field in an isolated condenser is $E_v$. ($b$) When filled with a dielectric the material is polarized and the field is reduced to $E$. ($c$) Relation between $E$ and $E_v$.

(a)     (b)     (c)

$$E_v = \frac{\sigma}{\varepsilon_0} \qquad E = \frac{\sigma - P}{\varepsilon_0} \qquad E = E_v - \frac{P}{\varepsilon_0}$$

† If we had considered a condenser maintained in a circuit at a constant potential we could show that the insertion of the dielectric involves a further flow of charge such that the total charge, for the same potential, is increased by a factor $\varepsilon$. Here again the free charges on the condenser plates are partly offset by the induced charges in the dielectric.

But from equation (12.3), $\sigma = \varepsilon\varepsilon_0 E$ and equation (12.4) becomes

$$\varepsilon\varepsilon_0 E - P = \varepsilon_0 E,$$

or

$$P = (\varepsilon - 1)\varepsilon_0 E. \tag{12.5}$$

Thus a dielectric which reduces the electrostatic field by a factor $\varepsilon$ produces a polarization $P$ which is $(\varepsilon - 1)\varepsilon_0$ times the real field. Equation (12.5) may also be written

$$(\varepsilon - 1) = \frac{P}{\varepsilon_0 \varepsilon} = \chi, \tag{12.6}$$

where $\chi$ is defined as the volume susceptibility of the dielectric. In what follows we shall show how $P$ can be derived in terms of atomic models. Before doing so, however, we refer for completeness to the displacement vector $D$.

### 12.1.2 *Displacement vector D*

In passing from vacuum to a dielectric the field changes from $E_v$ to $E$. Taking the simplest case of a parallel block of dielectric normal to the field we see from figure 99($c$) that

$$E = E_v - P/\varepsilon_0$$

or

$$E_v = E + P/\varepsilon_0 = E\varepsilon.$$

There is a discontinuity in the field. Another quantity called the displacement vector $D$ may now be defined

$$D = \varepsilon E. \tag{12.7}$$

Then in vacuum, $D_v = \varepsilon_v E_v = E_v$ (since $\varepsilon_v = 1$), and in the dielectric

$$D = \varepsilon E = E_v.$$

The quantity $D$ is thus continuous. We shall not use it further.

### 12.2 **Polarization of gases**

#### 12.2.1 *Mechanisms of polarization*

We first consider the behaviour of gases since the molecules are, generally, so far apart that interactions between them may be neglected. Consequently each gas molecule 'sees' the dielectric field $E$. This polarizes the molecule in three ways:

($a$) By distorting the electronic cloud around the nucleus or nuclei of each molecule; we call the polarization due to this mechanism $P_e$.

(*b*) By stretching or bending bonds in polar molecules; we call this $P_a$. Usually this is small $(P_a \sim 0.1 P_e)$. and can be neglected.

(*c*) By orienting existing molecular dipoles; we call this $P_d$.

We first consider the electronic polarization $P_e$. Suppose the molecule is initially non-polar. As a result of electronic displacement the individual molecule (or atom) will acquire a dipole moment $p$. As we shall see immediately $p$ is proportional to $E$, so we may write

$$p = \alpha E, \tag{12.8}$$

where $\alpha$ is the molecular (or atomic) polarizability. If there are $n$ molecules per m$^3$

$$P_e = np = n\alpha E. \tag{12.9}$$

We at once note that the dimensions of $P_e$ are charge/area whilst the dimensions of $E$ are charge/(distance)$^2 \varepsilon_0$, so they are dimensionally identical apart from the $1/\varepsilon_0$ term. Hence $n\alpha$ has dimensions of $\varepsilon_0$. Thus dimensionally

$$[\alpha] = \left[\frac{1}{n}\right]\varepsilon_0 = [L^3]\varepsilon_0.$$

In fact, as we shall see in what follows, $L$ is of the order of the molecular radius.

### 12.2.2 *The polarization of neutral monatomic molecules*

Consider a neutral atom such as argon or neon, which consists of a central positive charge $Ze$ and an electron cloud of charge $-Ze$ (see figure 100(*a*)). We assume that the electron cloud has uniform charge density $\rho$ and extends over a sphere of radius $r$. If a field $E$ is applied the

Figure 100. An atom of atomic number $Z$ considered as a central positive charge surrounded by a negative charge of uniform density. The application of a field $E$ displaces the positive relative to the negative charge.

electron cloud experiences a force $EZe$ in one direction and the nucleus an equal force in the opposite direction. The charges are displaced relative to one another until the attractive force between them exactly equals the force exerted by the field. Let the displacement be $x$ (see figure 100($b$)). The attraction between the nucleus and the electron cloud is the force between $Ze$ on the nucleus and that part of the electronic charge which is within the sphere of radius $x$.

The latter is $Ze(x/r)^3$. The attractive force (see figure 100($c$)) is then

$$f = \frac{(\text{positive charge}) \times (\text{negative charge})}{4\pi\varepsilon_0 x^2},$$

i.e.

$$f = Ze \times Ze\left(\frac{x}{r}\right)^3 \frac{1}{4\pi\varepsilon_0 x^2} = \frac{Z^2 e^2 x}{4\pi\varepsilon_0 r^3}. \qquad (12.10)$$

This is balanced by the distorting force

$$f = EZe. \qquad (12.11)$$

Equating (12.10) and (12.11) we have

$$E = \frac{1}{4\pi\varepsilon_0} \frac{Zex}{r^3}. \qquad (12.12)$$

But $Zex$ is the charge times the distance between the charges, i.e. it is the atomic dipole $p$. Hence

$$p = 4\pi\varepsilon_0 r^3 E. \qquad (12.13)$$

A similar result is obtained if the atom is assumed to be a perfectly conducting sphere of radius $r$. Hence

$$\alpha = p/E = 4\pi\varepsilon_0 r^3.$$

It is interesting to compare the radius of an atom or a molecule, deduced from transport processes (see Table 14, p. 72) with the value of $(\alpha/4\pi\varepsilon_0)^{\frac{1}{3}}$ derived from the measured dielectric properties ($p/E$) of the material. Typical results are given in Table 31.

Table 31. *Polarizability and molecular radius r (units $10^{-10}$ m)*

| Gas | $\alpha = p/E$ $(10^{-40}\,\text{F}\,\text{m}^2)$ | $(\alpha/4\pi\varepsilon_0)^{\frac{1}{3}}$ $(10^{-10}\,\text{m})$ | $r$ $(10^{-10}\,\text{m})$ |
|---|---|---|---|
| He | 0.24 | 0.6 | 1.1 |
| Ar | 1.9 | 1.2 | 1.7 |
| $H_2$ | 0.95 | 0.95 | 1.3 |
| $N_2$ | 1.9 | 1.2 | 1.8 |
| $CO_2$ | 3.1 | 1.4 | 2.1 |

Continuing from equation (12.13) we may now write

$$P_e = np = n4\pi\varepsilon_0 Er^3,$$

Combining this with equation (12.5) we have

$$\varepsilon - 1 = 4\pi nr^3. \tag{12.14}$$

### 12.2.3 *Dielectric properties and van der Waals critical data*

We now make a wide-ranging and rather unexpected correlation between $\varepsilon$ and critical gas data. We note that $4\pi nr^3$ is three times the volume of the molecules in $1\ \text{m}^3$ of gas. The critical volume of the gas is 12 times the volume of the molecules. It follows that $4\pi nr^3$ is equal to $\frac{1}{4}(\rho/\rho_c)$, where $\rho$ is the density of the gas and $\rho_c$ the density at the critical point. Hence

$$(\varepsilon - 1) = \tfrac{1}{4}\frac{\rho}{\rho_c}. \tag{12.15}$$

According to Maxwell's theory of electromagnetic waves, at the same wavelength, the refractive index $\mathscr{R}$ is related to the dielectric constant $\varepsilon$ and the magnetic permeability $\mu$ by the relation

$$\mathscr{R} = (\varepsilon\mu)^{\frac{1}{2}}. \tag{12.16}$$

Table 32. *Dielectric properties of gases at s.t.p.*

| 1 | 2 | 3 | 4 | 5 | 6 | 7 |
|---|---|---|---|---|---|---|
| | | $\rho$ | $\rho_c \times 10^{-3}$ | $\tfrac{1}{4}\dfrac{\rho}{\rho_c}$ | $(\varepsilon - 1)$ | $2(\mathscr{R} - 1)$ |
| Type | Gas | $(\text{kg m}^{-3})$ | $(\text{kg m}^{-3})$ | $\times 10^4$ | $\times 10^4$ | $\times 10^4$ |
| Inert | He | 0.18 | 0.069 | 6.5 | 0.7 | 0.7 |
| | Ne | 0.90 | 0.48 | 4.7 | 1.3 | 1.3 |
| | Ar | 1.78 | 0.53 | 8.4 | 5.5 | 5.7 |
| | Xe | 5.90 | 1.15 | 12.2 | 15.0 | 14.0 |
| Non-polar | $H_2$ | 0.09 | 0.031 | 7.2 | 2.7 | 2.8 |
| molecules | $N_2$ | 1.25 | 0.31 | 10.0 | 5.8 | 5.9 |
| | $O_2$ | 1.43 | 0.43 | 8.3 | 5.2 | 5.4 |
| | $Cl_2$ | 3.21 | 0.57 | 14.1 | — | 15.4 |
| | $CO_2$ | 1.98 | 0.46 | 10.8 | 9.9 | 9.9 |
| Molecules | $NH_3$ | 0.77 | 0.23 | 8.4 | 71 | 7.5 |
| with dipoles | $SO_2$ | 2.93 | 0.53 | 13.8 | 82 | 13.2 |
| | $H_2O$ | 0.60 | 0.32 | 4.7 | 80 | 5.0 |

For dielectrics $\mu$ is equal to unity to within one part in $10^6$. Since for gases $\varepsilon$ is only slightly greater than unity we may write

$$\mathcal{R} = \varepsilon^{\frac{1}{2}} = [1 + (\varepsilon - 1)]^{\frac{1}{2}} = 1 + \tfrac{1}{2}(\varepsilon - 1), \tag{12.17}$$

so that

$$(\mathcal{R} - 1) \approx \tfrac{1}{2}(\varepsilon - 1). \tag{12.18}$$

Combining with (12.15) we have

$$\tfrac{1}{4}\left(\frac{\rho}{\rho_c}\right) = (\varepsilon - 1) = 2(\mathcal{R} - 1). \tag{12.19}$$

These quantities have been compared in Table 32 for inert (monatomic) gases, non-polar polyatomic gases, and for molecules with dipoles. The dielectric constant is found from static or low-frequency measurements.

There are three clear conclusions. First, for the inert and non-polar molecules the agreement between $\tfrac{1}{4}(\rho/\rho_c)$ and the observed values of $(\varepsilon - 1)$, see columns 5 and 6, is reasonably good except for He where the inner two electrons are very tightly bound. We conclude that our calculation of the molecular polarization is valid for polyatomic molecules as well as for monatomic molecules. Secondly, for all these gases there is very good agreement between $(\varepsilon - 1)$ and $2(\mathcal{R} - 1)$ – columns 6 and 7. Since the optical refractive index is a measure of the interaction of the molecules with the high-frequency electric field associated with light waves we deduce that this is the same as the electronic polarization produced by a static or low-frequency electric field in these cases.

Thirdly, we note that polar molecules give good agreement between $\tfrac{1}{4}(\rho/\rho_c)$ and $2(\mathcal{R} - 1)$ – columns 5 and 7 – since both are a measure of high-frequency electronic polarization. There is poor agreement between $(\varepsilon - 1)$ and $2(\mathcal{R} - 1)$ – columns 6 and 7 – since $\varepsilon$ is a low-frequency determination and includes the orientation effects of the molecule. By contrast $\mathcal{R}$ involves optical frequencies where only the electronic polarizations can respond. Indeed the difference between $(\varepsilon - 1)$ and $2(\mathcal{R} - 1)$ for ammonia suggests that the orientation of the dipole contributes $63 \times 10^{-4}$ to the low-frequency dielectric constant.

## 12.3 Polarization of polar molecules

12.3.1 Apart from the electronic polarization, polar molecules show two other effects:

(a) The stretching or bending of bonds. This produces a polarization $P_a$ which is generally very small. We shall not attempt to estimate it quantitatively.

(*b*) An orientation of existing dipoles. The polarization $P_d$ due to this is large and may be as much as $100P_e$.

Materials in which existing dipoles are oriented by the applied field are sometimes referred to as para-electric materials. In what follows we provide a simplified model by means of which we can estimate the magnitude of $P_d$ quantitatively.

The simplest example of a dipole (see figure 101) is a diatomic molecule with a charge $e$ at one end and $-e$ at the other and a separation of say 0.1 nm. The permanent dipole has a moment

$$p = \text{charge} \times \text{separation}$$
$$= 1.6 \times 10^{-19} \times 10^{-10}$$
$$= 1.6 \times 10^{-29} \, \text{C m.} \tag{12.20}$$

In the absence of a field these dipoles are randomly oriented and show no resultant moment. If an electric field could orient all the molecules we should obtain a polarization per m$^3$ given by

$$P_d = np.$$

For a gas at s.t.p. $n$ is about $3 \times 10^{25}$ per m$^3$. Hence

$$P_d \approx 5 \times 10^{-4} \, \text{C m}^{-2}.$$

The field $E_d$ produced in the dielectric by these dipoles is

$$E_d = P_d/\varepsilon_0 \approx 6 \times 10^7 \, \text{V m}^{-1}. \tag{12.21}$$

This would be more than enough to ionize most gases, but experiments show that it is virtually impossible to ionize a gas in this way. The reason is that the calculation ignores thermal motion which vigorously opposes the lining-up of the dipoles. In most practical situations the polarization achieved is very much less than that calculated above.

### 12.3.2 *Orientation of dipoles in an electric field: the Langevin function (para-electric behaviour)*

We consider the effect of a field $E$ on an assembly of dipoles of moment $p$. The field attempts to orient the dipoles, thermal motion to maintain disorder.

We first calculate the energy of a single dipole in the field if its axis makes an angle $\theta$ to the direction of $E$ (see figure 102(*a*)). Let the dipole

Figure 101. A simple electrostatic dipole.

$10^{-10}$ m

consist of charges $q$ and $-q$ separated by a distance $2a$. Then $p = 2aq$. The potential energy is defined as work done on the system by external forces. The forces on the charges are $qE$ and $-qE$ (figure 102($b$)). If both charges were at some arbitrary position, say at O, we should do work $-qEx$ on the positive charge and $qE(-x)$ on the negative charge in moving them to their present positions. The total potential energy $\varepsilon$ is

$$\varepsilon = -2qEx = -2qEa \cos \theta$$

or

$$\varepsilon = -Ep \cos \theta. \tag{12.22}$$

We now consider how much space is occupied by orientations between $\theta$ and $\theta + d\theta$. Construct a sphere of arbitrary radius $R$ (figure 102($c$)). The solid angle subtended between $\theta$ and $\theta + d\theta$ is

$$d\omega = \frac{\text{area of annulus}}{R^2} = \frac{2\pi R \sin \theta \, R \, d\theta}{R^2},$$

or

$$d\omega = 2\pi \sin \theta \, d\theta. \tag{12.23}$$

We now apply the Boltzmann distribution. The number of dipoles with axes between $\theta$ and $\theta + d\theta$ is

$$dn = B \exp\left(\frac{-\varepsilon}{kT}\right) \times (\text{element of space})$$

$$= B \exp\left(\frac{pE \cos \theta}{kT}\right) 2\pi \sin \theta \, d\theta. \tag{12.24}$$

As this is a rather clumsy expression we make the following substitutions:

$$\left. \begin{array}{l} x = \dfrac{pE}{kT}, \\[2ex] z = \cos \theta \text{ so that } dz = -\sin \theta \, d\theta. \end{array} \right\} \tag{12.25}$$

Figure 102. The orientation of a dipole of moment $2aq$ in a field of strength $E$.

Then equation (12.24) becomes

$$dn = -2\pi B \exp(zx)\, dz. \tag{12.26}$$

We find $B$ by integrating $dn$ for $\theta$ from 0 to $\pi$, that is for $z$ from $+1$ to $-1$. This gives $n$, the total number of dipoles per m$^3$;

$$n - 2\pi B \int_{-1}^{+1} \exp(zx)\, dz. \tag{12.27}$$

(The change in sign for $n$ follows because we have reversed the order of the limits of integration.) Because of symmetry the resultant polarization normal to $E$ will be zero, and only the polarization parallel to $E$ will be additive. Each dipole contributes a dipole moment parallel to $E$ of amount $p \cos \theta$. Thus the resultant dipole moment per unit volume $P_d$ will be

$$P_d = \int_{\theta=0}^{\theta=\pi} p \cos \theta \, dn$$

$$= 2\pi B p \int_{-1}^{+1} z \exp(zx)\, dz. \tag{12.28}$$

We can express the ratio of $P_d$ to $n$ (using equations (12.27) and (12.28)) as:

$$\frac{P_d}{n} = \frac{p \int z \, e^{zx} \, dz}{\int e^{zx} \, dz}, \tag{12.29}$$

where the integral is from $z = -1$ to $z = +1$.

In most practical situations $zx$ is small so that

$$e^{zx} \approx 1 + zx. \tag{12.30}$$

Equation (12.29) then becomes

$$\frac{P_d}{n} = \left[ \frac{p \int z(1+zx)\, dz}{\int (1+zx)\, dz} \right]_{-1}^{+1} = p\frac{x}{3}. \tag{12.31}$$

Substituting from equation (12.25) for $x$ we have

$$P_d = \frac{np^2 E}{3kT}. \tag{12.32}$$

We may at once note that for most dielectrics $E$ cannot exceed about $10^7 \, \text{V m}^{-1}$, otherwise breakdown occurs. For a typical value of $p = 1.6 \times 10^{-29} \, \text{C m}$, using $k = 1.4 \times 10^{-23} \, \text{J K}^{-1}$, and $T = 300 \, \text{K}$, we have

$$\frac{pE}{kT} \approx 0.03,$$

so that our assumption implied in equation (12.30) is valid,

Exact integration of equation (12.29) gives the Langevin function

$$P_d = np\left[\coth\left(\frac{pE}{kT}\right) - \frac{kT}{pE}\right]. \tag{12.33}$$

This function is plotted in figure 103 and shows that under normal conditions the linear part of the curve, far removed from saturation, is the only part that is observable. The contribution of $P_d$ to the dielectric constant may now be calculated. We have

$$P_d = (\varepsilon - 1)_d \varepsilon_0 E = \frac{np^2 E}{3kT}$$

or

$$(\varepsilon - 1)_d = \frac{np^2}{\varepsilon_0 3kT}. \tag{12.34}$$

For ammonia at s.t.p. we have $n = 3 \times 10^{25}\,\mathrm{m}^{-3}$, $p \approx 0.5 \times 10^{-29}\,\mathrm{C\,m}$. Then

$$(\varepsilon - 1)_d = 67 \times 10^{-4}.$$

This agrees extremely well with the value quoted in the discussion on p. 287. Measurements of the dielectric constant are often used by the chemist to deduce the value of the molecular dipole.

The orientation polarization can, of course, be increased without exceeding the breakdown of the dielectric by reducing the temperature. This cannot, however, be carried very far since most *polar* substances solidify fairly readily and under these conditions molecular orientation becomes very difficult or impossible.

Figure 103. The Langevin relation for the polarization of dipoles produced by a field $E$.

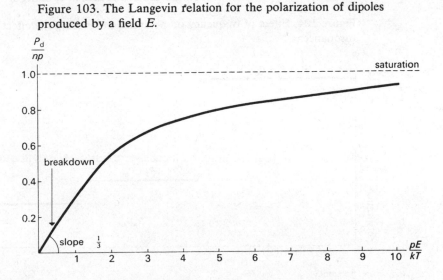

### 12.3.3 *Effect of frequency of electric field*

We can at once see the powerful effect of frequency on the polarization. We write

$$P = P_d + P_a + P_e$$

$$= \frac{np^2}{3kT} E + P_a + nr^3 4\pi\varepsilon_0 E.$$

In a static or low-frequency field all these processes will exert their full part. However, the orientation of the whole molecule is relatively slow, and as the frequency increases the orientation lags behind. When the frequency reaches a value of about $10^{12}$ the dipoles are unable to follow the oscillations of the field: only random orientations remain and these contribute nothing to the resultant polarization. Of the polarization, $P$, only $P_a$ and $P_e$ remain. At a somewhat higher frequency the stretching of the bonds becomes too sluggish and $P_a$ drops out. Only $P_e$ remains above a frequency of $10^{15}$. This is in the optical range. The behaviour is sketched in figure 104. However, this ignores the fact that both $P_a$ and $P_e$ may show resonance effects. In the next section we shall consider only $P_e$ and show that resonance occurs at a critical frequency and that this accounts for optical dispersion and anomalous dispersion.

### 12.4 **Optical dispersion and anomalous dispersion**

We now analyse the effect of an alternating electric field on the electronic polarization $P_e$; the treatment is similar to that given by G. Joos. For simplicity we consider a single electron, mass $m$, charge $-e$,

Figure 104. Effect of frequency on polarization ignoring the effect of resonance.

bound to its equilibrium position relative to its nucleus by a force of force-constant $k$. In free oscillation the equation of motion of the electron is

$$m\ddot{x} + kx = 0,$$

or

$$\ddot{x} + \omega_0^2 x = 0, \tag{12.35}$$

where $\omega_0$ is the natural angular frequency (radians $s^{-1}$) and $\omega_0^2 = k/m$. The natural frequency $\nu_0$ is given by

$$\nu_0 = \frac{\omega_0}{2\pi} \text{ Hz.} \tag{12.36}$$

We apply a sinusoidal electric field

$$E = E_0 \sin \omega t, \text{ frequency } \nu = \frac{\omega}{2\pi}. \tag{12.37}$$

Ignoring any 'frictional' effects the equation of motion is

$$\ddot{x} + \omega_0^2 x = \frac{-e}{m} E_0 \sin \omega t. \tag{12.38}$$

The steady-state solution is

$$x = \frac{-eE_0 \sin \omega t}{m} \frac{1}{\omega_0^2 - \omega^2}. \tag{12.39}$$

The dipole moment per atom is simply

$$p = -ex. \tag{12.40}$$

Figure 105. ($a$) For an undamped system, resonance at frequency $\nu_0$ produces a change in $\mathcal{R}$ from $+\infty$ to $-\infty$. ($b$) With a damped system the transition is sharp but the infinite values do not occur.

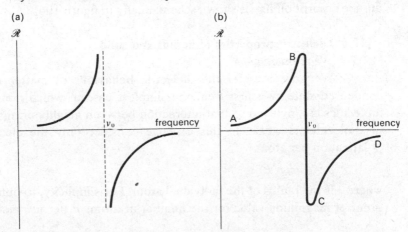

Hence

$$p = \frac{e^2 E_0 \sin \omega t}{m(\omega_0^2 - \omega^2)} = \frac{e^2}{4\pi^2 m(\nu_0^2 - \nu^2)} E. \tag{12.41}$$

The volume polarization $P_e$ is simply $n$ times this. Using the relationship

$$(\varepsilon - 1) = 2(\mathcal{R} - 1) = \frac{P_e}{\varepsilon_0 E} = \frac{np}{\varepsilon_0 E}$$

and substituting for $p$ from equation (12.41) we obtain

$$\mathcal{R} = 1 + \frac{ne^2}{8\pi^2 \varepsilon_0 m(\nu_0^2 - \nu^2)}. \tag{12.42}$$

This is plotted in figure 105(a). It is seen that approaching $\nu_0$ the refractive index rises to plus infinity and just beyond $\nu_0$ it falls to minus infinity. This is the absorption frequency of the atom and is generally of the order of $10^{15}$. In practice, radiation resulting from the oscillation involves a certain amount of damping or 'friction', and this prevents infinite values of $\mathcal{R}$ from being reached. The main features are now shown in figure 105(b). The refractive index increases with frequency over the region AB; this is the region of 'normal' dispersion. The sudden change in behaviour along CD was, at one stage, considered anomalous and was so called. The above analysis shows, however, that 'anomalous' dispersion is just as normal as 'normal' dispersion.

The frequency $\nu_0$ at which electronic resonance occurs is the same as that involved in the derivation of van der Waals forces. For this reason these forces are sometimes called 'dispersion' forces (see pp. 18–20).

Similarly the frequency at which a resonance effect occurs in $P_a$ is of the order of $10^{12}$, so dispersion also occurs in the infrared region. The combined behaviour for a dipolar molecule with a single electronic and atomic absorption line is shown schematically in figure 106.

## 12.5 Dielectric properties of liquids and solids
### 12.5.1 *Weak interaction*

We now consider the dielectric behaviour of matter in the condensed state. We first treat the simplest case in which non-polar molecules are involved so that interaction between neighbouring atoms or molecules is weak. In this case we consider only the electronic polarization per atom

$$p = 4\pi \varepsilon_0 r^3 E, \tag{12.43}$$

where $r$ is the radius of the individual atom. For simplicity, to obtain an order of magnitude value for the number of atoms $n$ per m$^3$, assume a

simple cubic structure. Then

$$n = \left(\frac{1}{2r}\right)^3.$$ (12.44)

The polarization per unit volume, using equations (12.43) and (12.44), is

$$P = np = \frac{\pi \varepsilon_0}{2} E.$$ (12.45)

Hence

$$\varepsilon - 1 = \frac{P}{\varepsilon_0 E} = \frac{\pi}{2}.$$ (12.46)

Consequently

$$\varepsilon = 1 + \frac{\pi}{2} \approx 2.6,$$

$$\mathcal{R} = \varepsilon^{\frac{1}{2}} \approx 1.6.$$ (12.47)

This analysis shows that for many simple liquids and solids the dielectric constant will be of order 2–3 and the refractive index 1.4–1.7. Clearly in a crystal which has well-defined structural anisotropy, so that the polarizability depends on crystal direction, there will be a corresponding dependence of $\varepsilon$ and $\mathcal{R}$ on crystal direction. This is the basic cause of birefringence.

Figure 106. Effect of frequency on polarization of a dielectric allowing for resonance effects.

### 12.5.2 *Strong interaction: the Clausius–Mossotti equation*

When atoms or molecules are far apart as in a gas, they 'see' the average dielectric field $E$. When, however, they are close together as in a liquid or solid, the atomic (or molecular) dipoles may themselves interact. For example if polarization produces a local dipole of strength $p = 4\pi\varepsilon_0 r^3 E$, the field produced by this dipole at a distance $r$ is of order of magnitude $p/4\pi\varepsilon_0 r^3$, i.e. about $E$; i.e. the field produced at a neighbour by an induced dipole may be about as large as the field itself. This increases the polarization effect.

We have now to consider how to estimate the resultant field acting on each molecule. The treatment is due to Lorentz. We choose a molecule O and construct a sphere of radius $R$ large compared with the molecular separation (see figure 107). The resultant effect at O is the sum of ($a$) the field produced by the material within $R$ and ($b$) the field produced by the material outside $R$. The latter can be treated as a continuum since it is so far away from O. The field due to this is the same as the equivalent surface charges over the spherical surface. The calculation is as follows.

Construct an annulus on the surface of the cavity making an angle $\theta$, $\theta + \mathrm{d}\theta$ with the direction of the polarization field. The surface charge density is

$$-P \cos \theta. \tag{12.48}$$

The charge on the annulus is

$$-P \cos \theta \, 2\pi R \sin \theta R \, \mathrm{d}\theta. \tag{12.49}$$

Figure 107. Charge distribution in a spherical cavity within a dielectric.

This produces a coulomb field at O pointing along OA. The vertical components cancel out, the horizontal ones are additive. The horizontal component at O, for the annulus, is

$$\left(\frac{1}{4\pi\varepsilon_0}\right)P\cos\theta\,2\pi R\sin\theta R\,d\theta\frac{1}{R^2}\cos\theta$$
$$= (1/2\varepsilon_0)P\cos^2\theta\sin\theta\,d\theta. \tag{12.50}$$

For the whole sphere, for $\theta$ ranging from 0 to $\pi$, the net field $E'$ at O due to the surface charge is

$$E' = \frac{P}{2\varepsilon_0}\int_0^\pi\cos^2\theta\sin\theta\,d\theta = \frac{P}{2\varepsilon_0}\left[\frac{-\cos^3\theta}{3}\right]_0^\pi = \frac{P}{3\varepsilon_0}. \tag{12.51}$$

The resultant field at O is thus $E + E'$ or

$$E + \frac{P}{3\varepsilon_0}. \tag{12.52}$$

We now have to consider the contribution $(a)$ due to material within the sphere. Lorentz showed that for a cubic array of dipoles or for a random array of dipoles as in a liquid or a glass, the local field is zero. For such materials we may, therefore, write

$$E_{\text{local}} = E + \frac{P}{3\varepsilon_0}. \tag{12.53}$$

The induced dipole per molecule is

$$p = \alpha E_{\text{local}}, \tag{12.54}$$

therefore

$$P = np = n\alpha\left(E + \frac{P}{3\varepsilon_0}\right). \tag{12.55}$$

Hence

$$\frac{P}{E} = \frac{n\alpha}{1 - n\alpha/3\varepsilon_0}. \tag{12.56}$$

The basic relation between $P$ and the average field $E$ in the dielectric is

$$(\varepsilon - 1) = \frac{P}{\varepsilon_0 E} = \frac{n\alpha}{\varepsilon_0 - n\alpha/3}. \tag{12.57}$$

Hence

$$(\varepsilon + 2) = (\varepsilon - 1) + 3 = \frac{3\varepsilon_0}{\varepsilon_0 - n\alpha/3}. \tag{12.58}$$

Taking the ratio of equations (12.57) and (12.58), we have finally

$$\frac{\varepsilon - 1}{\varepsilon + 2} = \frac{n\alpha}{3\varepsilon_0}. \tag{12.59}$$

This relation is fairly well obeyed for polar liquids and for dilute solutions of dipolar molecules in a non-polar solvent, as well as for ionic solids of cubic structure. It is known as the Clausius–Mossotti equation.

Before we leave this section we may form some idea of the importance of interaction compared with the results obtained if interaction is ignored. In the latter case, as we saw above,

$$\varepsilon - 1 = \frac{P}{\varepsilon_0 E} = \frac{n\alpha}{\varepsilon_0}. \tag{12.60}$$

Consider three cases:

(a) Weak interaction, say $\varepsilon \approx 1$; we see that equation (12.60) is identical with equation (12.59).

(b) Slight interaction, say $\varepsilon = 3$; equation (12.59) then gives a value for $(\varepsilon - 1)$, which is $\frac{5}{3}$ times bigger than that given by equation (12.60).

(c) Large dipole interaction, say $\varepsilon = 70$; equation (12.59) gives a value for $(\varepsilon - 1)$ which is 24 times larger than that given by equation (12.60).

There is yet a further type of interaction which we have not discussed here, but shall refer to briefly in the next chapter. This is the analogue of

Figure 108. Variation of dielectric constant with frequency, for water at 20 °C and for ice at −5 °C and −40 °C. The large dielectric constant at low frequencies at −40 °C shows that molecular rotation is still possible (see p. 213).

ferromagnetism where interaction is intrinsically so large that even without the presence of an applied field the material is strongly polarized. It would correspond to a situation (see equation (12.58)) where $n\alpha \approx 3\varepsilon_0$ so that $\varepsilon \to \infty$. Such materials are known as ferroelectrics. A ferroelectric material such as barium titanate has a dielectric constant of over 1000.

### 12.5.3 *Effect of frequency*

With matter in the condensed state frequency has the same effect on the dielectric constant as it has on gases. However, the dipole polarization in liquids is usually more sluggish than in gases – with solids it is often completely absent since rotation of the whole molecular unit within the lattice is difficult or impossible, particularly at temperatures well below the melting point. Results for water and ice are shown in figure 108.

### 12.6 **Summary**

We may summarize the ideas described in this chapter in the following way.

1. The dielectric properties of matter arise from the polarization produced by an electrostatic field. Polarization occurs in three ways:

(*a*) Distortion of electron clouds; all atoms (or molecules) show this effect.

(*b*) With polar molecules stretching or bending of bonds can produce a small contribution to the polarization.

(*c*) With polar molecules orientation of the molecule may produce a very large polarization: the field has to compete with thermal randomization.

2. Temperature has no effect on (*a*) since this involves electronic properties. Increasing temperature resists the polarizing influence of the field for both (*b*) and (*c*). It is particularly marked with (*c*). However, the molecular orientation cannot be increased indefinitely by lowering the temperature since orientation of the molecular dipoles may become impossible when the material is solid.

3. Frequency of the electrostatic field has a marked effect on all types of polarization. Orientation becomes increasingly difficult as the frequency is raised and becomes impossible at frequencies above about $10^{12}$ Hz. The stretching or bending of bonds also becomes unable to follow the field at frequencies above about $10^{14}$ Hz, and only the electronic polarization remains up to a frequency of $10^{15}$–$10^{16}$ Hz. This is in the optical range and is therefore the only part of the dielectric

mechanism which plays a part in the refractive index of the material. There is usually a resonance effect at frequencies in this region and this accounts for optical dispersion. The bond stretching (or bending) mechanism also shows a resonance effect but this generally occurs in the infrared.

4. With matter in the condensed phase the behaviour is complicated by neighbour–neighbour interaction. This can greatly increase the effective polarization of the material.

# 13

---

## *Magnetic properties of matter*

### 13.1 Introduction

We know that there are certain metallic bodies which attract one another and which, if suitably suspended will tend to point in the north–south direction. These materials are magnetic. We also know that if a current is passed through a coil of wire, the coil itself behaves like a magnet and will attract a bar magnet or another similar coil. If we insert certain 'magnetic' materials into the coils the interaction between the coils will be changed. The coil also possesses a certain self-inductance and if we insert similar 'magnetic' materials into the coil its self-inductance will be increased.

These are some of the basic observations involved in magnetism and the purpose of this chapter is to discuss the magnetic properties of matter in terms of atomic and electronic mechanisms. There are many resemblances to dielectric effects the main differences being that

(a) there is no such thing as an isolated magnetic pole;

(b) there is no magnetic equivalent of a condenser;

(c) there is no dielectric analogue of a solenoid.

The most fundamental approach to magnetism is due to Einstein. He considered the electric field generated by a moving charge and the force it would exert on another charge moving parallel to it with the same velocity. Because of relativity effects the space ahead of the moving charge is crowded and this compression of space in which the electric field operates produces an additional force on the other charge over and above that calculated from the classical coulombic law. This extra force turns out to be identical with the magnetic force exerted between two parallel wires carrying an electric current.

Thus magnetism is a second-order relativistic electrostatic effect – it is not a new type of force. We shall not discuss this in detail any further.

## 13.2 **The magnetic equations**

We first define the magnetic field $B$ as that type of field which produces a force on a conductor carrying a current. If a field $B$ (teslas) acts at an angle $\theta$ on a short length of conductor $dl$ (metres) carrying a current $I$ (amperes) it is found that the conductor experiences a force $F$ (newtons) in a direction at right angles to a plane containing $B$ and $dl$ (figure 109(a)) of magnitude

$$F = BI\ dl\ \sin\ \theta. \tag{13.1}$$

Consider now a coil of irregular shape in the $xy$ plane, the field $B$ pointing in the $x$ direction (figure 109($b$)). There is a force on the element at $P$ of amount

$$BI\ dl\ \sin\ \theta = BI\ dy \tag{13.2}$$

out of the paper and an equal force at $Q$ into the paper. These two forces exert a couple

$$BI\ dy\ x_1 = BI\ dA, \tag{13.3}$$

where $x_1$ is the distance PQ and $dA$ is the area of the strip between P and Q. For the whole coil the couple is then $BIA$, where $A$ is the area of the coil. Thus the coil behaves as though it had a magnetic moment $IA$. We may thus define magnetic moment $m$ of a coil as

$$m = IA \tag{13.4}$$

and add to this that when lying in a coplanar field $B$ it experiences a couple

$$G = BIA = Bm. \tag{13.5}$$

Figure 109. ($a$) A field $B$ acting on a current element $I$ produces a force normal to the plane containing $B$ and $I$. ($b$) Calculation of moment exerted by a field $B$ on a coil carrying current $I$.

(a)            (b)

Thus a coil carrying a current behaves in a magnetic field like a magnet. It is for this reason that ever since A. M. Ampère, it has been considered that the magnetism of atoms is due to circulating currents within the atom; consequently every 'north' pole has associated with it an equal 'south' pole. For this reason it is not possible to envisage the existence of isolated magnetic poles in the way that there are isolated charges in electrostatics. The earlier view regarded the circulating currents as arising from electrons travelling in orbits; later work has included the electron spin which, from this point of view, can also be regarded as a minute circulating current.

Let us now consider the behaviour of a long solenoid containing $n_0$ turns per unit length. The solenoid is assumed to be so long that the free ends do not produce any disturbing effects at the centre. If we were to place a small test coil at the centre of the solenoid we should find that the solenoid produces a couple on the coil from which we should deduce that the magnetic field has a value

$$B = \mu_0 I n_0, \tag{13.6}$$

where $\mu_0$ = permeability of free space = $4\pi \times 10^{-7}\,\mathrm{H\,m^{-1}}$.

Suppose the solenoid was of length $l$ and has a cross-section of area $A$. We could compare its magnetic properties with a permanent magnet of length $l$, cross-sectional area $A$, possessing a uniform magnetic moment $M$ per unit volume. The total moment of the magnet is

$$M \times \text{volume} = MAl. \tag{13.7}$$

For the solenoid each turn produces a magnetic moment (see equation (13.4))

$$m = IA,$$

so that the moment of the whole solenoid is

$$m \times n_0 l = IAn_0 l. \tag{13.8}$$

Comparing equations (13.7) and (13.8) we see that the coil and the magnet have identical magnetic properties if

$$MAl = IAn_0 l,$$

i.e.

$$M = In_0. \tag{13.9}$$

Since the field in the solenoid (see equation (13.6)) is $\mu_0 I n_0$, it follows that the equivalent field inside the magnet is also $\mu_0 I n_0$. Consequently, from equation (13.9)

$$B = \mu I n_0 = \mu_0 M. \tag{13.10}$$

We now consider the effect of filling the solenoid with 'magnetic' material. We assume that the solenoid is so long that the end poles have a negligible 'demagnetizing' effect. Alternatively this could be achieved by forming the solenoid in a torroidal shape so that there are no free ends. The field in the solenoid in air is (see equation (13.6))

$$B = \mu_0 I n_0.$$

Self-induction experiments show that $B$ is increased by the presence of the 'magnetic' material by a factor $\mu$ known as the magnetic permeability; which for weakly magnetic material is a constant (for strongly magnetic materials, $\mu$ is itself a function of $B$). Thus within the solenoid

$$B = \mu \mu_0 I n_0.$$

But we may also regard $B$ as being the sum of two parts: one part is due to the solenoid itself ($\mu_0 I n_0$) and the other part is due to the magnetism $M$ produced in the magnetic material ($\mu_0 M$), thus we may write:

$$B = \mu \mu_0 I n_0 = \underset{\substack{\downarrow \\ \text{due to} \\ \text{solenoid}}}{\mu_0 I n_0} + \underset{\substack{\downarrow \\ \text{due to} \\ \text{magnetic} \\ \text{material}}}{\mu_0 M}. \tag{13.11}$$

We now define $I n_0$ as the magnetizing field $H$ of the solenoid (amperes per metre)

$$H = I n_0. \tag{13.12}$$

Then equation (13.11) can be written as

$$B = \mu \mu_0 H = \mu_0 H + \mu_0 M, \tag{13.13}$$

$$\mu - 1 = M/H = \chi, \tag{13.14}$$

where $\chi$ is the volume susceptibility of the magnetic material.

If there are $n$ atomic or molecular or electronic units per unit volume each of which acquires magnetic moment $m$ we have

$$M = nm.$$

Then equation (13.14) becomes

$$\mu - 1 = \frac{nm}{H} = n\alpha, \tag{13.15}$$

where $\alpha$ is defined as the magnetic polarizability per unit (atom, molecule, electron).

In experimental magnetism we apply a magnetizing field $H$ and find the magnetic moment $M$ produced in the material under examination. In what follows we shall show how we may deduce $M$ or $\chi$ or $\alpha$ in terms of atomic mechanisms. Before doing so we may draw on the parallelism between a solenoid and a magnet to emphasize a further point of difference between magnetic and dielectric properties. If electric charges are placed on a dielectric the electrostatic field is always from positive to negative whether it is observed inside or outside the dielectric. With a permanent magnet, however, the situation is different. Outside the material the field is in the direction away from the 'north' pole and towards the 'south' pole: within the material the field is from south to north. The explanation for this is quite obvious if we replace the permanent magnet by an equivalent solenoid. Perhaps more relevant is the idea that every elementary magnetic unit within the magnet, down to the individual atom or even the spinning electron, can be regarded as equivalent to a minute circulating current.

### 13.3  Diamagnetism: Langevin's treatment

All atoms possess a diamagnetic component. This is because an applied field always modifies the circulating 'electronic' current in such a way that it opposes the field. We consider the simplest case, an electron of charge $e$, mass $m_e$ moving with velocity $v$ in a circular orbit of radius $r$. If $\tau$ is the time for the electron to describe one revolution, $e$ behaves like a current $I$ of magnitude

$$I = \frac{e}{\tau} = \frac{ev}{2\pi r}.$$  (13.16)

The magnetic moment of the orbiting electron is

$$m = \pi r^2 I = \frac{evr}{2}.$$  (13.17)

Consider now the effect of a field of induction $B$ which is normal to the orbit. The flux $\phi$ in the circuit is

$$\phi = \pi r^2 B.$$  (13.18)

If the field varies with time an e.m.f. will be induced in the circuit of magnitude

$$\text{e.m.f.} = -\frac{d\phi}{dt} = -\pi r^2 \frac{dB}{dt}.$$  (13.19)

This e.m.f. acts around the current loop and is equivalent to an electrostatic field $E$, where

$$\text{e.m.f.} = 2\pi r E$$
$$= -\pi r^2 \frac{dB}{dt}.$$  (13.20)

Hence

$$E = -\frac{r}{2}\frac{dB}{dt}. \tag{13.21}$$

This field exerts a force $eE$ on the electron and this in turn produces a change $dv$ in its velocity:

$$eE = m_e\frac{dv}{dt} = -\frac{er}{2}\frac{dB}{dt}, \tag{13.22}$$

$$\frac{dv}{dt} = \frac{er}{2m_e}\frac{dB}{dt}. \tag{13.23}$$

Consequently in the time that $B$ changes by $\Delta B$, $v$ changes by $\Delta v$ where

$$\Delta v = -\frac{er}{2m_e}\Delta B. \tag{13.24}$$

According to equation (13.17) a change in $v$ produces a corresponding change in the magnetic moment of the orbit:

$$\Delta m = \frac{er}{2}\Delta v.$$

Combining with equation (13.24),

$$\Delta m = -\frac{e^2r^2}{4m_e}\Delta B. \tag{13.25}$$

We see that $\Delta m$ is always in a sense which opposes $\Delta B$. Consider, for example, two identical orbits with opposite orbital momenta. Each orbit has a magnetic moment $m$ but the net moment is zero. If a field $\Delta B$ is applied one orbit is slowed down and the other is speeded up: the net change is additive for the two orbits (see figure 110).

Figure 110. Sketch showing nature of diamagnetism; (*a*) when no field is applied the net amount is zero, (*b*) when field $\Delta B$ is applied there is an additive change in the magnetic moments of both orbits.

(a)                                    (b)

There is another point of interest. If the orbital velocity is reduced by $\Delta v$, the centrifugal force on the electron is reduced by $\Delta(m_e v^2/r) = 2m_e v \Delta v/r$. From equation (13.24) this equals $ve \Delta B$. But the field $\Delta B$ itself exerts a radial force on the electron of amount: velocity $\times$ charge $\times \Delta B = ve \Delta B$. This exactly compensates for the reduced centrifugal force: consequently if the initial orbit is stable under some specified force between nucleus and electron the diamagnetic mechanism does not introduce any change.

For diamagnetic materials $B \approx H$ to 1 part in $10^5$. Equation (13.25) then becomes

$$\Delta m = \frac{-e^2 r^2}{4m_e} \Delta H. \tag{13.26}$$

By definition the magnetic susceptibility per electron is simply $\Delta m/\Delta H$. If there are $n$ atoms per unit volume the susceptibility per unit volume is

$$\chi = n\frac{\Delta m}{\Delta H} = -\frac{n}{4}\sum\frac{e^2 r^2}{m_e}, \tag{13.27}$$

where the summation is for all the orbital electrons in each atom.

If the orbits are not normal to $B$ each orbit undergoes precession and $\chi$ is slightly decreased. Similarly if the orbits are not circular a lower value of $\chi$ is obtained. A better value for $\chi$ is

$$\chi = -\frac{n}{6}\sum\frac{e^2 r^2}{m_e}. \tag{13.28}$$

We may at once derive an order-of-magnitude value for $\chi$. If one mole of a typical metal occupies $10^{-5}$ m$^3$, $n_0 = 6 \times 10^{28}$. If there are say 10 orbital electrons and $r$ is of order $10^{-10}$ m we have

$$\chi = -(2 \text{ to } 3) \times 10^{-6}. \tag{13.29}$$

Consequently $\mu_0 M$ is quite small compared with $B$. This value of $\chi$ is typical for most solids and liquids. In the gaseous state the main change is in $n_0$. At s.t.p. $n_0$ is about $10^3$ times smaller than in the condensed state so that for gases at s.t.p.

$$\chi \approx 10^{-9}.$$

Note that $\chi$ is independent of temperature since the internal structure of the atom is virtually unchanged by normal temperatures.

In the model described above, diamagnetism is attributed to the influence of the magnetic field on the orbital momentum of the electrons. What then happens to the s-electrons, which possess zero orbital momentum? A detailed quantum theory treatment shows that s-electrons possess zero angular momentum only in the absence of a magnetic field. Upon the application of a magnetic field they acquire a small amount of orbital momentum. This produces an equivalent amount of diamagnetism to that calculated above. Thus it follows that all the electrons contribute to diamagnetism.

## 13.4 Paramagnetism: the Langevin function

If an atom possesses a permanent magnetic dipole because the electrons have either orbital momentum or spin, the dipole will orient in an applied magnetic field and produce a magnetization in the same direction as the applied field. Materials which are made of such atoms are attracted (weakly) by a magnetic field and are known as paramagnetic substances. As in the case of dielectric dipoles the orientation by the applied field is opposed by thermal disorder.

We first consider the order of magnitude of the permanent dipole – arising from an orbiting electron. For a circular orbit of radius $r$ and electron velocity $v$ we have (see equation (13.17)).

$$m = \frac{evr}{2}.$$
(13.30)

But the angular momentum $m_e vr$ is quantized in units of $h/2\pi$, where $h$ is Planck's constant. Putting

$$m_e vr = l\left(\frac{h}{2\pi}\right),†$$
(13.31)

where $l$ is an integer, and substituting in equation (13.30) we have

$$m = l\left(\frac{eh}{4\pi m_e}\right).$$
(13.32)

The quantity in the bracket is known as the Bohr magneton. As a first approximation we expect to find certain atoms in which the angular momenta do not cancel out, so that one orbit (or one spin) is left unpaired. The magnetic moment of such an atom will be one Bohr magneton, the magnitude of which is

$$\frac{eh}{4\pi m_e} \approx 10^{-23} \text{ J T}^{-1}.$$
(13.33)

Since orbits, and spins, tend to occur in pairs of opposite momenta (or spins) we should not expect to find atoms with magnetic moments much

† More accurately, as mentioned in Chapter 1 we should write, in place of $l$, $[l(l+1)]^{\frac{1}{2}}$.

greater than this. This has an important consequence which greatly eases further analysis. As we shall see below, at low temperatures and moderate fields $H$, it is possible to orient all the dipoles and produce magnetic saturation. Nevertheless, even at this stage, the magnetic moment $M$ per unit volume is small compared with $H$. Consequently the difference between $B$ and $\mu_0 H$, or between $H$ and the true field seen by the individual atom, is negligible and can be ignored. In what follows, therefore, we consider simply the effect of the field $B$.

Following the analysis for para-electric behaviour given in the previous chapter, we expect to find for small and moderate fields a volume susceptibility

$$\chi = \frac{M}{H} \approx \frac{\mu_0 M}{B} = \frac{\mu_0 n m^2}{3kT}. \tag{13.34}$$

For a solid occupying $10^{-5} \text{ m}^3$ per mole, $n = 6 \times 10^{28}$. Assuming $m$ is one magneton we find that, at 300 K, $\chi$ has a value

$$\chi \approx 6 \times 10^{-4}.$$

This is 50–100 times greater than the diamagnetic susceptibility. Consequently the diamagnetic susceptibility, which all atoms possess, is swamped by the paramagnetic susceptibility if the atom contains an unpaired orbit or spin.

For strong fields or low temperatures we obtain the corresponding function

$$\chi = \frac{nm}{H} \left( \coth \frac{mB}{kT} - \frac{kT}{mB} \right). \tag{13.35}$$

This is drawn in figure 111. Although this curve is identical with that obtained for para-electrics there are two basic differences between paramagnetic and para-electric materials. The first concerns saturation. Low temperatures clearly favour saturation in both cases. However, most dielectrics possessing electrostatic dipoles become solids at low temperature and orientation becomes difficult or impossible; orientation is thus frozen-out. With paramagnetics we are only concerned with the orientation of orbits (or spins) and this can occur at the very lowest temperatures. Thus one can greatly enhance paramagnetism by operating at very low temperatures; one cannot in general do this with para-electrics.

As an example of saturation with paramagnetics we note from figure 111 that this occurs when $mB/kT > 5$, i.e. when

$$B \approx \mu_0 H > \frac{5kT}{m}. \tag{13.36}$$

If $M = 10^{-23}$ J T$^{-1}$ and $T = 1$ K, we obtain saturation when

$H > 6 \times 10^6$ A m$^{-1}$.

Such a magnetizing field can be easily obtained and, with it, saturation can be achieved at temperatures of about 1 K. It should be noted that this argument is valid for gases where the individual molecules are far apart ($n$ relatively small) and do not interact with one another. It also applies to ionic salts where each ion behaves like an isolated dipole but in such materials the number of dipoles per unit volume will not be large. For example in chrome alum, $CrK(SO_4)_2 \cdot 12H_2O$, the $Cr^{3+}$ ions have permanent dipoles and so contribute to the paramagnetism, but the $K^+$ and $SO_4^-$ ions show only diamagnetic effects. The number of $Cr^{3+}$ ions per unit volume is relatively small. Thus for both gases and ionic salts, although saturation can be obtained as indicated above with reasonable magnetizing fields by working at low temperatures the overall magnetization will not be as large as might at first sight be expected because $n$ will not be very large.

On the other hand with paramagnetic metals, where $n$ is large, the argument given above is not valid. The metal may be considered to consist of positive ions in a sea of electrons. The positive ions generally consist of closed electronic shells or an even number of electrons with paired spins. They are diamagnetic. The valency electrons (which provide electrical conductivity) have spins which can be aligned by an applied magnetizing field (see section 13.6 below). These therefore contribute to

Figure 111. The Langevin function for paramagnetism, treating the orientation as a continuous (classical) phenomenon.

the paramagnetism of the metal. The field also affects the translational motion of these electrons and it can be shown that this produces a diamagnetic contribution which is just about one-third of the paramagnetism. We shall not deal with this but continue discussion of the paramagnetic effect. Since this is due to the free valency electrons in the metal, the number able to take part in this process is not the total number of free electrons $n$ but, as pointed out on p. 226, only those at the top of the Fermi level $E_F$, i.e. only a fraction of order $kT/E_F$. Consequently the quantity $n$ per unit volume in equation (13.34) should be replaced by $nkT/E_F$. The volume susceptibility them becomes

$$\chi = \mu_0 \frac{nkT}{E_F} \frac{m^2}{3kT} = \frac{\mu_0 nm^2}{3E_F}. \tag{13.37}$$

We see that the susceptibility is independent of temperature, a result in good agreement with the observed behaviour of paramagnetic *metals.*

We have discussed the possibility of saturation with paramagnetic salts and gases. The second difference between paramagnetic and dielectric behaviour concerns the nature of the polarizing field. Consider the typical dense solid discussed after equation (13.34): $n = 6 \times 10^{28}$ per m$^3$, $m$ is of the order of 1 Bohr magneton. Suppose that at very low temperatures ($\sim 1$ K) all the dipoles are aligned by the application of a magnetizing field $H = 6 \times 10^6$ A m$^{-1}$ as suggested by equation (13.36). Then the magnetic moment per unit volume $M = nm = 6 \times 10^5$. Thus $M$ is still a good deal smaller than $H$ and we do not need to distinguish between $B = \mu \mu_0 H$ and $B = \mu_0 H + \mu_0 M$ (see equation (13.13)). Thus we are not faced with the difficult problem that occurs in dielectric studies of attempting to allow for the effect of the polarization itself on the field producing the polarization. Only in ferromagnetism does this become important. In that case, as we shall see below, the interaction due to the magnetic polarization of the material is so strong that it completely dominates the behaviour, i.e. the resultant field is dominated by $M$ rather than by $H$.

The difference between the minute volume susceptibility of paramagnetics as compared with para-electrics is due to the difference in size of the unit dipole. In electrostatics $p$ has a value of about $1.6 \times 10^{-29}$ C m: in paramagnetism $m$ is of the order of the Bohr magneton and has a value of about $10^{-23}$ J T$^{-1}$. We have:

$$\text{in para-electricity } \chi = \frac{np^2}{\varepsilon_0 3kT}, \text{ in paramagnetism } \chi = \frac{\mu_0 nm^2}{3kT}.$$

Thus for materials with the same values of $n$ the ratio $\chi_{mag}/\chi_{diel}$ is equal to

$$\frac{\mu_0\varepsilon_0 m^2}{p^2} \approx 10^{-5}.$$

Note that the quantity $\mu_0\varepsilon_0$ in the numerator is equal to the reciprocal of the square of the velocity of light. This emphasizes the basic connection between electrostatic and electromagnetic phenomena deduced by Clerk Maxwell and explained by Einstein in terms of relativistic effects (see p. 301). It is for the latter reason that magnetic phenomena are 'second-order' effects compared with electrostatic phenomena. Thus for matter in the condensed state $(\mu - 1)$ is of the order $10^{-4}$ in magnetism while $(\varepsilon - 1)$ is of order 1 to 10 in dielectrics.

### 13.5 Ferromagnetism

#### 13.5.1 *Theory of ferromagnetism*

Certain materials such as iron, nickel and cobalt show very high magnetization $M$ (typically $M \approx 10^6$ A m$^{-1}$) for fairly modest fields even at room temperature. They reach magnetic saturation very easily. With an 'ordinary' specimen of iron this occurs for magnetizing fields of the order of a few $10^{-1}$ A m$^{-1}$ (see figure 112); with good single crystals in favourable orientations for a field of order only $10^{-4}$ A m$^{-1}$, so under these conditions $\chi$ is about 1000.

A typical value of $M$ at saturation is about $10^6$ A m$^{-1}$. If one mole occupies $10^{-5}$ m$^3$ the number of atoms per m$^3$ is $6 \times 10^{28}$. At saturation, if these give a moment of $10^6$ A m$^{-1}$ each atom must contribute a magnetic

Figure 112. Magnetic moment $M$ produced in a typical magnetic material for magnetizing fields of order $10^{-1}$ A m$^{-1}$.

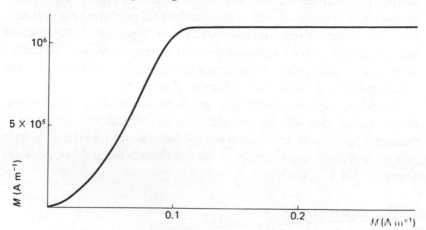

moment of $10^6/(6 \times 10^{28}) \approx 1.6 \times 10^{-23}$, i.e. about 1 Bohr magneton. This suggests that a saturated ferromagnetic consists of completely aligned dipoles. In this sense they are like paramagnetics, but with paramagnetics this result can only be achieved by using enormous fields or by operating at extremely low temperatures. With ferromagnetics alignment occurs with relatively weak fields.

It was this conclusion that led A. Weiss to suggest that these materials are spontaneously magnetized over fairly large domains and that these domains are easily aligned when an external field is applied. There is a cooperative interaction between the atoms so that an additional field $\lambda M$ acts on each atom. This resembles the Lorentz field in dielectrics $\frac{4}{3}\pi P$, but the quantity $\lambda$ is immensely larger, of the order $10^4$. The resultant field acting on each atom and producing magnetization may then be written

$$B = \mu_0(H + \lambda M). \tag{13.38}$$

The Langevin equation for the alignment of the dipoles becomes

$$M = mn\left(\coth x - \frac{1}{x}\right), \tag{13.39}$$

where now

$$x = \frac{m}{kT}(H + \lambda M)\mu_0. \tag{13.40}$$

The behaviour of the Langevin equation as a function of $H$ is best seen by rewriting equation (13.39) as

$$\frac{M}{nm} = \coth x - \frac{1}{x}, \tag{13.41}$$

Figure 113. Graphical method of explaining ferromagnetism. (a) Graph of equation (13.41). (b) Graph of equation (13.42). (c) Superposition of both graphs. For curve (i) there is no intersection, i.e. ferromagnetism does not occur. For curve (ii) intersection implies that ferromagnetism does occur.

and equation (13.40) as

$$\frac{M}{nm} = \frac{xkT}{m^2\lambda n\mu_0} - \frac{H}{\lambda nm}.$$                               (13.42)

Consider the behaviour when $H = 0$. This corresponds to the spontaneous magnetization of ferromagnetic materials in the absence of an applied field. We plot equation (13.41) as $M/nm$ against $x$ and obtain the typical curve of figure 113($a$). The slope near the origin is $\frac{1}{3}$. In figure 113($b$) we have plotted equation (13.42) as $M/nm$ against $x$, for $H = 0$. We obtain a straight line of slope $kT/m^2\lambda n\mu_0$. Clearly if this slope is greater than $\frac{1}{3}$ no solution is possible and there is no magnetization (figure 113($c$), curve (i)). If the slope is less than $\frac{1}{3}$, magnetization and saturation are easily achieved (figure 113 ($c$), curve (ii)).

It is at once evident that there is a critical temperature above which no ferromagnetization will occur. This is known as the Curie temperature $T_c$ and is given by

$$\frac{kT_c}{m^2\lambda n\mu_0} = \frac{1}{3}.$$                                       (13.43)

If the saturation value of $M$ is written as $M_0$ ($M_0 = nm$) the drop-off of $M$ with increasing temperature is shown in figure 114 as a plot of $M/M_0$ against $T/T_c$. Above $T_c$ only paramagnetism and diamagnetism remain. In what follows we shall ignore the diamagnetism and calculate in a very simple way the paramagnetism above $T_c$. If the fields are not too strong

Figure 114. Drop in magnetic moment of a ferromagnetic material as temperature is raised. Above the Curie temperature $T_c$ the material shows only diamagnetism and paramagnetism.

the Langevin equation (13.41) may be written

$$\frac{M}{nm} = \frac{x}{3}. \tag{13.44}$$

Similarly, using equation (13.43), equation (13.42) may be written

$$\frac{M}{nm} = \frac{xT}{3T_c} - \frac{H}{\lambda nm}. \tag{13.45}$$

Substituting for $x$ from equation (13.44) we obtain

$$\frac{M}{nm}\left(\frac{T}{T_c} - 1\right) = \frac{H}{\lambda nm}. \tag{13.46}$$

Consequently the paramagnetic susceptibility becomes

$$\chi = \frac{M}{H} = \frac{1}{\lambda}\left(\frac{T_c}{T - T_c}\right). \tag{13.47}$$

From equation (13.43) this becomes

$$\chi = \frac{nm^2\mu_0}{3k(T - T_c)}. \tag{13.48}$$

We see at once that $\lambda$ has disappeared explicitly although, of course, it is implicit in the value of $T_c$. Equation (13.48) is well obeyed and gives good quantitative agreement for most ferromagnetic materials on the assumption that each atom contributes of the order of 1 or 2 Bohr magnetons.

Finally we may note that in the ferromagnetic range an applied field has little effect compared with the internal field. Even for a field $H \approx 1$ A m$^{-1}$ the interval $mH\mu_0/kT$ shown in figure 115 is small and has very little effect on the resultant magnetization.

Figure 115. An applied field $H$ has little effect on the magnetization of a ferromagnetic material.

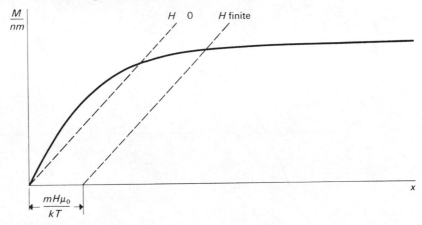

### 13.5.2 *The origin of the internal field: ferromagnetism and anti-ferro-magnetism*

The internal field $B_0$ at saturation in a ferromagnetic material is given by

$$B_0 = \lambda M \mu_0 = \lambda m n \mu_0.$$

For typical values $\lambda = 5000$, $m = 1$ Bohr magneton ($10^{-23}$ J T$^{-1}$). $n = 6 \times 10^{28}$ atom m$^{-3}$, $\mu_0 = 4\pi \times 10^{-7}$ Hm$^{-1}$ we obtain $B_0 = 4000$ T. This is an extremely strong field, very much greater than the ordinary field acting on a dipole due to the magnetization of its neighbours (for this $\lambda$ would be of order unity as in the dielectric case).

It is not possible to explain the origin of this large internal field in simple terms. Indeed it introduces very difficult quantum mechanical problems. We may, however, describe it in the following terms. The large value of $\lambda$ arises from the 'exchange' interaction between the atomic dipoles. This interaction depends on the ratio of the orbital radius $r$ to the distance $a$ between neighbouring atoms. Over a critical range ($a/r$ about 3) the magnetic exchange interaction is large and positive. This implies that for atoms with large dipoles and of the correct $a/r$ ratio strong positive interaction will occur, $\lambda$ will be large and the material will be ferromagnetic (Fe, Co, Ni). This is shown schematically in figure 116. It is also seen that for small values of $a/r$ the interaction can be negative. This favours an antiparallel orientation of neighbouring spins, and materials showing this property (MnF$_2$, MnO, CoO) are known as anti-ferromagnetic substances: at low temperatures the interaction is effective and an applied field produces little magnetization so that the susceptibility is

Figure 116. Exchange interaction in ferromagnetic materials as a function of $a/r$.

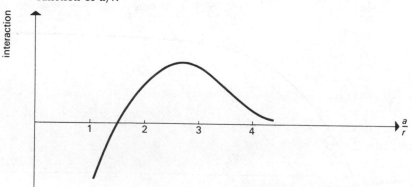

small. As the temperature is raised the anti-ferromagnetic interaction becomes less effective and the susceptibility of the material increases. Above a critical temperature known as the Néel temperature the interaction is completely swamped by thermal motion. The material ceases to be anti-ferromagnetic and becomes paramagnetic in the, more or less, classical sense. The detailed behaviour is very complicated and we shall not discuss it further.

### 13.6 Quantum treatment of magnetic properties

The classical treatment of diamagnetism is satisfactory from two points of view. First, the concept of induced currents producing fields which oppose the applied field seems reasonable and tractable. Secondly, it gives the right answer for the diamagnetic susceptibility. From all other points of view it is wrong.

Detailed considerations, first described by N. Bohr in 1911, show that if classical mechanics are applied *rigorously* to an orbiting electron the paramagnetic and diamagnetic terms cancel one another exactly so that there is no net induced magnetization. The reason is briefly as follows. Assuming that an electron orbiting around a nucleus is a stable classical system, the radius of a given orbit can have a continuous range of values from 0 to ∞ rather than the one particular size assumed in the Langevin theory. Further, whereas the Langevin theory applies the Boltzmann energy distribution to the external degrees of freedom of the orbiting electron, i.e. to the orientation of the orbits in paramagnetism, it does not do so to the internal degrees of freedom, i.e. to the orbital energies of the electron in its treatment of diamagnetism. This is logically inconsistent. If the Boltzmann distribution is applied to both orientation energies and orbital energies the paramagnetic and diamagnetic parts exactly compensate one another. The Langevin treatment gives the correct answer because it assumes a finite radius and an orbital energy independent of temperature; in a sense it incorporates hidden quantum conditions although in fact Langevin's theory (1905) preceded Bohr's quantized atom (1913) by eight years.

Classical electromagnetic concepts cannot really be applied to a classical atom since a non-quantized orbiting electron constitutes an unstable system (see below). The only situation for which such concepts can be applied meaningfully is to a free 'electron-gas', for example the electron plasma in a metal. Here again, the effect of a magnetic field is to produce diamagnetic and paramagnetic effects which exactly cancel one another. The reason is simply that in classical electromagnetism a field can deflect an electron but it cannot do work on it, so that although the electrons are deflected into helical paths their energy distribution remains unaltered.† This implies that the magnetic field produces no energy change in the system, i.e. the system behaves as though it has no net magnetic property. A discussion of this issue is given in a very readable and interesting form in *The Theory of Electric and Magnetic Susceptibilities*.‡

As mentioned above, the classical atom is unstable; the orbiting electrons would spiral into the nucleus radiating all their energy in the process. The effective lifetime of such an atom would be infinitesimally small. Atoms exist as stable entities precisely because energy states (or angular momenta) are quantized. To produce a perturbation of the energies by the classical methods is quite invalid. However, quantum mechanics shows that there is a genuine diamagnetic effect which can be calculated rigorously. The analysis is far too advanced to be given here. It gives the same answer as the classical treatment which we have given in this chapter in section 14.2. For this reason, as in most other elementary texts, we have persisted in giving the simple treatment: the answer is right though the physics is not entirely satisfactory.

We have taken over from our chapter on dielectric properties the classical treatment of paramagnetism. With para-electrics quantum effects play a trivial part and the classical treatment is perfectly satisfactory. This is not so with paramagnetics. The reason is that the individual dipole can take up only a discrete number of orientations. If, for example, an atomic system has a resultant angular momentum represented by a vector of length $\frac{3}{2}$, the only orientations which the system can acquire in a magnetic field correspond to components of $\frac{3}{2}$, $\frac{1}{2}$, $-\frac{1}{2}$ and $-\frac{3}{2}$ in the direction of the field. In classical physics the orientations would cover the whole continuous range such that the components have values from $+\frac{3}{2}$ to $-\frac{3}{2}$. The effect on paramagnetism is not very marked but it is worth considering the simplest possible case when the dipole of the individual atom is due to a single electron spin. The only orientations allowed are those which are exactly parallel and antiparallel to the applied field. If there are $n$ atoms per $m^3$, and of these $n_+$ are those with orientations along the field and $n_-$ those with orientations opposing the

---

† Of course a magnetic field exerts a force on a conductor through which a current is passing. If the conductor moves, work is done, but the energy does not come from the magnetic field. The energy is provided by the electric current flowing through the conductor in overcoming the back e.m.f. generated in the conductor itself as a result of its movement in the magnetic field.

‡ J. H. van Vleck, *The Theory of Electric and Magnetic Susceptibilities* (Clarendon Press, 1932).

field, we have by the Boltzmann principle

$$n_+ = A \exp\left(\frac{mB}{kT}\right) \tag{13.49}$$

$$n_- = A \exp\left(\frac{-mB}{kT}\right). \tag{13.50}$$

Since $n_+ + n_- = n$ we find that

$$A = n\left[\exp\left(\frac{mB}{kT}\right) + \exp\left(-\frac{mB}{kT}\right)\right]^{-1}. \tag{13.51}$$

The net moment per unit volume is then

$$M = (n_+ - n_-)m \tag{13.52}$$

$$= nm\frac{\left[\exp\left(\frac{mB}{kT}\right) - \exp\left(-\frac{mB}{kT}\right)\right]}{\left[\exp\left(\frac{mB}{kT}\right) + \exp\left(\frac{mB}{kT}\right)\right]}$$

$$= nm \tanh\left(\frac{mB}{kT}\right). \tag{13.53}$$

In form this curve is very similar to the Langevin function. However, there is a quantitative difference in the region of small fields. Since for small values of $x$, $\tanh x \approx x$ we have from equation (13.53) for small values of $H$,

$$M \approx nm\left(\frac{mB}{kT}\right) = \frac{nm^2B}{kT}.$$

Figure 117. Magnetization of a single crystal of iron along a cube direction [100] and along a face diagonal direction [110].

The susceptibility is then

$$\chi = \frac{M}{H} = \frac{nm^2}{kT}\frac{B}{H} = \frac{nm^2\mu\mu_0 H}{kTH} \approx \frac{\mu_0 nm^2}{kT}$$

since $\mu$ is very nearly equal to unity. Comparison with equation (13.34) shows that this is just three times the value deduced from the Langevin function. Similar corrections may be applied to the discussion of ferromagnetism and the Curie temperature given above. However the correction factor of 3 will only apply rigorously for a single-electron-spin system. With most materials the situation is usually more complicated and in fact the correction factor becomes less than 3.

### 13.7 Ferromagnetic domains

We consider here briefly the magnetization of a single crystal of a ferromagnetic material such as iron. With iron the 6 cube edge directions [100] are the directions of easy magnetization. If a magnetic field $H$ is applied along the edge direction the domains are aligned easily and saturation is achieved for a very small field. If the field is applied along, say, the face diagonal direction [110] the easy orientation directions are at 45° to the field. Consequently for small fields the net magnetization in the direction of $H$ is $1/\sqrt{2}$ of what it would be if the field were applied along the [100] direction. Experiments show that this is so (see figure 117). If the field is increased further magnetization can occur only by twisting the domain directions out of their easy orientations. This is a

Figure 118. Magnetization of material showing a single direction of easy magnetization. ($a$) Initial unmagnetized state. ($b$) Effect of applying a weak field to the right is to orient all the domains in that direction. This gives magnetic saturation; on removing the field the magnetization remains unchanged so that the remanence is almost equal to the saturation magnetization. ($c$) Magnetization curve showing that the remanence is very large and that demagnetization can be achieved only by reversing the sign of the field.

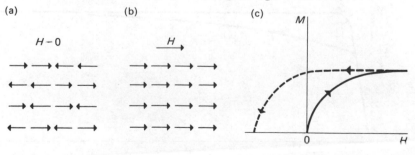

far more difficult process and relatively strong fields must be used to achieve it.

### 13.8 Magnetic hysteresis

When a magnetic field is applied to a magnetic material, the magnetization generally increases as the field is increased. If the field is reduced, the magnetization diminishes but does not follow the original curve – it 'lags behind'. This effect is known as magnetic hysteresis.

For the following simple description of magnetic hysteresis I am indebted to Dr Shoenberg. Consider the behaviour of a material which has a single easy direction of magnetization. In the unmagnetized state as many domains are magnetized to the left as to the right (see figure 118(*a*)). If a field is applied to the right the domains all become magnetized to the right and the material is saturated (figure 118(*b*)). The work done is used in moving the domain walls surrounding the right-pointing domains, until they swallow up the left-pointing domains. If the field is now removed no further change occurs and the material shows 100 per cent remanence (see figure 118(*c*)). The magnetization can be reduced only by applying a strong field to the left.

In general, metal specimens contain internal stresses. If the stresses are small they merely decide which of the easy directions of magnetization are easiest of all. If the stresses are large they may completely dominate the behaviour; in that case the easy domain directions are determined by the stresses rather than by crystal orientation. Consider for simplicity a specimen in which, as a result of local stresses, the domain directions are either horizontal, vertical or at 45° to these directions (figure 119(*a*)). If a

Figure 119. Magnetization of a material showing a limited number of directions of easy magnetization (vertical, horizontal and at 45°). (*a*) Initial unmagnetized state, (*b*) a horizontal field *H* aligns all the domains, (*c*) on removing the field the domains revert to their original directions but the sense is that of the field *H*. Under these conditions the remanence is about half the saturation magnetization.

strong horizontal field is applied all the domains line up with the field and the material achieves saturation (figure 119(*b*)). If the field is now removed the domains return to the original orientations they had in figure 119(*a*), but with the sense imposed on them by the applied field. This is shown in figure 119(*c*). Under these conditions the material has a remanance of about half the saturation magnetization.

### 13.9 Summary: comparison of dielectric and magnetic properties

We now summarize in note form the main points of resemblance and difference in dielectric and magnetic properties.

| Dielectric | Magnetic |
| --- | --- |
| Electronic polarization. Distortion of electron cloud. Critical frequency about $10^{15}$. Induced dipole: $$p = 4\pi\varepsilon_0 \, r^3 \, E$$ Always in direction of $E$ | Diamagnetism. Deceleration or acceleration or precession of those electrons possessing orbital momentum. Susceptibility per electron-mass $m_e$ $$-\tfrac{1}{6}\frac{e^2 \, r^2}{m_e}$$ Always opposes field $H$. |
| Unaffected by temperature | |
| Atomic polarization; stretching or bending of bonds if molecule has permanent dipole. Critical frequency about $10^{12}$ | None |
| Para-electric: orientation of molecular dipole $p$ by applied field $E$, opposed by thermal motion. Typical value of $p = 1.6 \times 10^{-29}$ C m. For small $E$, susceptibility per molecule is: $$\frac{n}{\varepsilon_0}\frac{p^2}{3kT}$$ | Paramagnetic: orientation of atomic dipole by applied field $H$, opposed by thermal motion. Typical value of $m = 10^{-23}$ J T$^{-1}$. For small $H$, susceptibility per atom is: $$\frac{\mu_0 n m^2}{3kT} \text{ (classical)}$$ $$\frac{\mu_0 n m^2}{kT} \text{ (quantum mechanical)}$$ |

Para-electric polarization is of order $10^4$ larger than paramagnetic. Hence need to consider real field rather than applied field. Very markedly dependent on temperature. Full behaviour described by Langevin function.

Para-electric behaviour cannot be easily enhanced by low temperatures because dielectric solidifies and molecules cannot be oriented.

Paramagnetic effects can be greatly enhanced by low temperatures since only orbits or spins have to be oriented.

*Ferroelectric*          *Ferromagnetic*

Cooperative phenomena in which internal field is augmented by some type of specific interaction. Large oriented domains of permanent dipoles. Above some critical temperature (Curie temperature) the interaction is ineffective, and the materials cease to be ferroelectric or ferromagnetic.

## *Appendix:* **Values of some physical constants**

| Quantity | Usual symbol | Value |
|---|---|---|
| Charge on electron | $e$ | $1.602 \times 10^{-19}$ C |
| Rest-mass of electron | $m_e$ | $9.110 \times 10^{-31}$ kg |
| Electron charge/mass | $e/m_e$ | $1.759 \times 10^{11}$ C kg$^{-1}$ |
| Rest-mass of proton | $M$ | $1.673 \times 10^{-27}$ kg |
| Ratio: proton/electron mass | $M/m_e$ | 1836 |
| Boltzmann constant | $k$ | $1.381 \times 10^{-23}$ J K$^{-1}$ |
| Avogadro's number | $N_A$ | $6.022 \times 10^{23}$ mol$^{-1}$ |
| Gas constant | $R$ | 8.314 J mol$^{-1}$ K$^{-1}$ |
| Mechanical equivalent of heat | — | 4.186 J cal$^{-1}$ |
| Volume of 1 mole of gas at s.t.p. | $V$ | $22.41 \times 10^{-3}$ m$^3$ |
| Faraday constant | $F$ | 96 490 C mol$^{-1}$ |
| Planck's constant | $h$ | $6.626 \times 10^{-34}$ J s |
| Planck's constant/$2\pi$ | $\hbar$ | $1.055 \times 10^{-34}$ J s |
| Bohr magneton [$eh/4\pi m_e$] | $\mu_B$ | $9.274 \times 10^{-24}$ J T$^{-1}$ |
| Bohr radius | $a_0$ | $0.529 \times^{-10}$ m |
| Gravitational constant | $G$ | $6.673 \times 10^{-11}$ N m$^2$ kg$^{-2}$ |
| Energy equivalent of 1 eV | | $1.602 \times 10^{-19}$ J |
| Velocity of light | $c$ | 299 800 km s$^{-1}$ |
| Permeability of free space | $\mu_0$ | $4\pi \times 10^{-7}$ H m$^{-1}$ |
| Permittivity of free space | $\varepsilon_0$ $= (\mu_0 c^2)^{-1}$ | $8.854 \times 10^{-12}$ F m$^{-1}$ |

The values in the above table for $N_A$, $R$, $J$ and $F$ are constants based on the number of atoms in 0.012 kg of $^{12}$C. The calorie in this table is the 15 °C calorie defined on p. 31.

# Index